Natural Gas Processing

Natural Gas Processing

Editor

Sanjay Khanwelkar

Natural Gas Processing

Edited by **Sanjay Khanwelkar**

Printed in 2017

ISBN: 978-1-68117-419-8

Library of Congress Control Number: 2015936537

Contents

Preface

Natural-gas processing begins at the well head. The composition of the raw natural gas extracted from producing wells depends on the type, depth, and location of the underground deposit and the geology of the area. Oil and natural gas are often found together in the same reservoir. The natural gas produced from oil wells is generally classified as associated-dissolved, meaning that the natural gas is associated with or dissolved in crude oil. Natural gas production absent any association with crude oil is classified as "non-associated." In 2009, 89 percent of U.S. wellhead production of natural gas was non-associated. Natural-gas processing is a complex industrial process designed to clean raw natural gas by separating impurities and various non-methane hydrocarbons and fluids to produce what is known as pipeline quality dry natural gas.

Editor

Life Cycle Water Use for Electricity Generation: A Review and Harmonization of Literature Estimates

J Meldrum[1], S Nettles-Anderson[2], G Heath[2], and J Macknick[2]

[1]Western Water Assessment and Institute of Behavioral Science, University of Colorado at Boulder, Boulder, CO 80309-0483, USA

[2]National Renewable Energy Laboratory, Golden, CO 80401-3305, USA

ABSTRACT

This article provides consolidated estimates of water withdrawal and water consumption for the full life cycle of selected electricity generating technologies, which includes component manufacturing, fuel acquisition, processing, and transport, and power plant operation and decommissioning. Estimates were gathered through a broad search of publicly available sources, screened for quality and relevance, and harmonized for methodological differences. Published estimates vary substantially, due in part to differences in production pathways, in defined boundaries, and in performance parameters. Despite limitations to available data, we find that: water used for cooling of thermoelectric power plants dominates the life cycle water use in most cases; the coal, natural gas, and nuclear fuel cycles require substantial water per megawatt-hour in most cases; and, a substantial proportion of life cycle water use per megawatt-hour is required for the manufacturing and construction of concentrating solar, geothermal, photovoltaic, and wind power facilities. On the basis of the best available evidence for the evaluated technologies, total life cycle water use appears lowest for electricity generated by photovoltaics and wind, and highest for thermoelectric generation technologies. This report provides the foundation for conducting water use impact assessments of the power sector while also identifying gaps in data that could guide future research.

INTRODUCTION

Water requirements throughout the electricity generation life cycle have important implications for the electricity sector. Although the operations of thermoelectric power plants require substantial water withdrawals (Kenny *et al* 2009, Macknick *et al* 2012), all electricity generation technologies, including those that do not require cooling for steam cycle processes, utilize water throughout their life cycles. This means that the power sector can be vulnerable to constraints caused by drought conditions and other changes in water resources not only directly, due to water required for operations (e.g., Huertas 2007, NETL 2009b), but also indirectly, due to water required throughout fuel supply chains and power plant equipment life cycles. For example, the 2011 drought in Texas not only impacted actual and expected statewide generation (Saathoff 2011) but also led to temporary shutdowns of hydraulic fracturing and other natural gas extraction operations (Carroll 2011, Passwaters 2011). Despite relatively low operational water demands compared to other generation technologies, photovoltaic and wind generation technologies also require water during manufacturing and construction.

The water requirements associated with choices along the life cycle of electricity generation, such as the selection of fuel type or cooling technology, are not well understood. Previous studies address water use across this life cycle to varying degrees (e.g., Gleick 1994, DOE 2006, Fthenakis and Kim 2010, Mielke *et al* 2010, McMahon and Price 2011), but these studies do not provide sufficient information for water and energy managers to quantify the magnitude and duration of expected impacts. Some reviews focus on water consumption and omit withdrawals for many or all processes (e.g., Gleick 1994, Mielke *et al* 2010, McMahon and Price 2011, Grubert *et al* 2012). Most reviews rely on a subset of available data. For example, two representative policy reviews (Mielke *et al* 2010, Wilson *et al* 2012) incorporate small amounts of new data but pull primarily from a few key, decades-old compilations (DOE 1983, Gleick 1994), as reflected in figure 1's schematic of the data provenance. Other reviews cover only specific life cycle stages (e.g., Macknick *et al* (2012) focus only on operations) or geographic contexts (e.g. Grubert *et al* (2012) evaluate life cycle water impacts from switching from coal- to natural gas-fired electricity generation in Texas). In addition, recent research (GAO 2009, Averyt *et al* 2013) questions the reliability of some data in commonly referenced statistics on thermoelectric power plant water use from the United States Geological Survey (USGS) (e.g., Solley *et al* 1998, Hutson *et al* 2004, Kenny *et al* 2009) and from the Energy Information Administration (EIA) (e.g., EIA 2011a, 2011b).

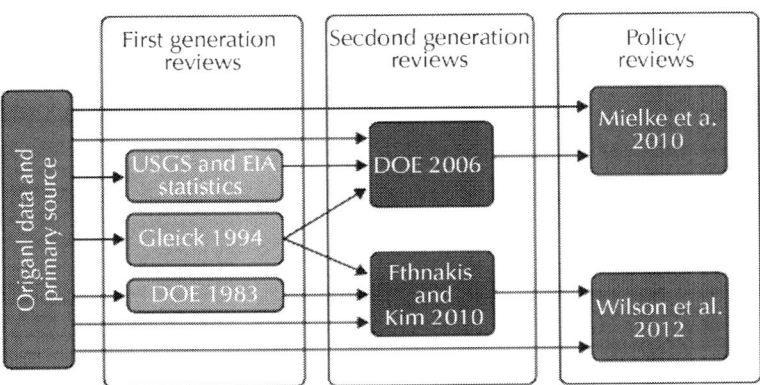

Figure 1: This schematic depicts the main provenance of data for some of the key references addressing the water use of electricity generation technologies. As shown, the sources for most data in even contemporary references are many decades old.

This paper assesses the life cycle water withdrawals and consumptive uses for renewable and non-renewable electricity generation technologies, hereafter referred to as *generation technologies*. It provides concise estimates of water use throughout the life cycle of electricity generation in the United States that are supported by a thorough review of available literature. Where available data differentiate water use among generation technology sub-categories, cooling technologies, or fuel cycle characteristics, we define each collective combination as a *production pathway* and present separate estimates for each. We collect estimates from a broad review of literature across multiple disciplines, publication types, and stages of the life cycle, applying consistent quality and relevance screens and harmonizing methodological differences. This review is intended to support more informed decisions on technological choice, research and development, and water resources management and to identify gaps where further research could significantly improve understanding of the water requirements for electricity generation.

SCOPE AND METHODS

Following the USGS (Kenny *et al* 2009), this paper classifies water use into water *withdrawals*, referring to 'water removed from the ground or diverted from a surface-water source for use' (p 49), and water *consumption*, referring to the portion of withdrawn water not returned to the 'immediate water environment' (p 47). The generic term water *use* is employed only where both metrics are discussed concurrently. Consistent with these definitions, we only address externally sourced water; therefore, we omit estimates addressing geologically produced and storm water that requires treatment. We focus only on water quantity requirements, although the life cycle of electricity generation may be associated with water quality issues as well (e.g., Lustgarten 2009, Ward2010, EPA 2011).

This paper reviews the water use throughout the life cycle of each of seven categories of electricity generation technologies: coal, natural gas, nuclear, concentrating solar power (CSP), geothermal, photovoltaics (PV), and wind. It does not address other technologies due to data complexities and uncertainties. For example, a separate literature addresses biopower's water requirements (e.g., Berndes 2008, Gerbens-

Leenes *et al* 2009, Stone *et al* 2010), but estimates vary by multiple orders of magnitude across regions, crops, and production methods. Reservoir evaporation complicates hydropower's accounting, leading to estimates ranging from zero to 18 000 gal MWh^{-1} (Gleick 1994, Torcellini *et al* 2003). Co-generation leads to challenges in allocating water use across co-products, and other generation technologies, such as ocean power, lack usable data.

Our broad literature review starts with the more than 2000 references amassed by the National Renewable Energy Laboratory's (NREL) Life Cycle Assessment (LCA) Harmonization project (www. nrel.gov/harmonization). Using keyword database searches, works cited lists, and known reference repositories, we expand this review to also include other peer-reviewed scientific literature, government reports and statistics, and corporate sustainability reports pertaining to life cycle water use or water use in any specific phase of the life cycle of electricity generation for the selected technologies.

We perform a series of three screens, described in more detail in the supplementary data (available atstacks.iop.org/ERL/8/015031/mmedia): two at the reference level and a third at the level of individual estimates. First, we screen this complete body of literature for any written quantification of water use within any electricity generation life cycle stage. Passing references receive a second screen, analogous to the screens for systematic review developed by the LCA Harmonization project (Moomaw *et al* 2011, Heath and Mann 2012), based on methods quality, completeness of reporting, and current technological relevance; this requires sufficient documentation by which the methods for developing results could be trusted and traced. With the exception of a few frequently cited, older sources (DOE 1983, Tolba 1985, Gleick 1994), which are retained because of both their importance to other literature and difficulties in tracking down many of their sources, we also eliminate references that did not provide primary data.

We gather data from all references passing the first two screens and present all non-duplicate estimates in the supplementary data (available at stacks.iop.org/ERL/8/015031/mmedia). In addition to removing duplicates, the third screen focuses on the reasonableness of individual estimates, considering both engineering principles and the preponderance of evidence. With a bias toward retaining estimates, we subject questionable estimates to further scrutiny, considering the thoroughness of documentation and the age of both the questionable reference and of alternative estimates' sources. We also omit otherwise reasonable estimates that lack sufficient disaggregation along the production pathway. In the results below, we discuss any unique estimates omitted in this third screen and otherwise focus presentation and analysis on data that pass all screens. Of the 138 sources passing the reference-level screens, one (Inhaber 2004) provides quantified water use data for all seven technologies we address, one more (Gleick 1994) addresses all but wind (for which water use is listed as 'negligible'), but the majority cover only one or two technologies each with primary data. A given reference can have multiple estimates, even for the same generation technology.

We categorize gathered data by generation technology and life cycle stage. As shown in figure 2, we separate the life cycle into three main stages: *fuel cycle*, which pertains only to coal, natural gas, and nuclear generation technologies; *power plant*, which represents the life cycle of the physical power plant equipment; and *operations*, which includes cooling for thermal technologies and all other plant operation and maintenance functions. Careful tracking of stage definitions and boundaries, which vary by study, is required to avoid double counting as much as possible when adding estimates across stages. This analysis does not account for electricity transmission, distribution, or end use, neither in terms of resource uses nor electricity losses.

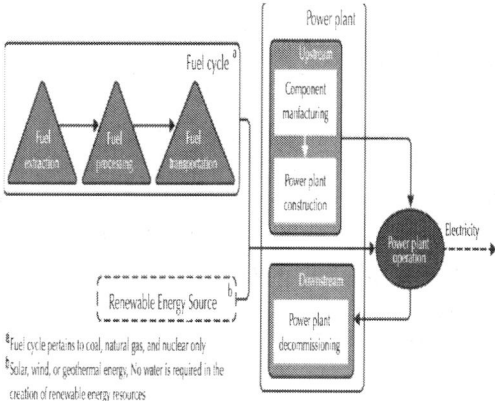

Figure 2: A schematic of the significant life cycle stages for each electricity generation technology demonstrates the additional role of fuel cycle wa-

Life Cycle Water Use for Electricity Generation: A Review and...

7

ter use in contributing to the life cycle water use for coal, natural gas, and nuclear generation technologies. The power plant life cycle stage consists of an upstream component manufacturing and plant construction phase and a downstream phase when the power plant is decommissioned.

Reflecting the spatial and temporal impacts of water use, most of this analysis focuses on water use disaggregated across life cycle stages. In addition, although the distinction between withdrawal and consumption is essential for understanding these impacts, many references do not specify which type of water use they report. Therefore, we infer this distinction from context and other information where possible and, where not, omit estimates because of insufficient reporting quality. In some cases of insufficient information, noted below, we assume withdrawals equal consumption, seeking balance between underestimating withdrawals when estimates actually report consumption and overestimating consumption when estimates actually report withdrawals.

We define life cycle water use factors (water$_{LC}$) as ratios of life cycle (LC) water use per unit of generated electricity, expressed as gallons per megawatt-hour (gal MWh^{-1}). We calculate factors for the life cycle water consumption and withdrawal associated with each generation technology. These factors represent weighted sums of the water use factors for each of the three major life cycle stages defined in figure 2:

$$water_{LC} = water_{FC} * \left(\frac{fuel_{lifetime}}{e_{lifetime}} \right)$$

$$+ water_{PP} * \left(\frac{1}{e_{lifetime}} \right) + water_{OP}$$

(1)

where water$_{FC}$ is the amount of water used in the fuel cycle (FC) per unit of fuel (expressed as gal ton^{-1} for coal, gal MMscf^{-1} for natural gas, and gal kg^{-1} converted, enriched, and fabricated uranium fuel (i.e., UO$_2$) for nuclear); e$_{lifetime}$ is the amount of electricity generated by a power plant over its lifetime (MWh/lifetime); fuel$_{lifetime}$ is the amount of fuel used by a power plant over its lifetime (ton/lifetime, MMscf/lifetime, or kg/lifetime, as appropriate); water$_{PP}$ is the amount of water used for component manufacturing, power plant construction, and power plant decommissioning (i.e. the power plant equipment life cycle (PP) as defined in figure 2) (gal/lifetime); and water$_{OP}$ is the amount of water used in the operations (OP) of the power plant per unit of generated electricity (gal MWh^{-1}).

In addition to water use estimates, we record parameters relevant to fuel attributes, fuel cycle characteristics, and power plant performance, which can influence the amount of water used in life cycle stages. Where possible with available data, we harmonize all estimates to the common performance parameters shown in table 1; the goal of harmonization is to reduce analytical variability by adjusting previously published estimates to ones based on a more consistent set of methods and assumptions (Heath and Mann 2012). These parameters are selected to match those published in LCA harmonization and other benchmark studies on electricity generated by coal (MIT 2007, Whitaker *et al* 2012), natural gas (EIA 2011c, O'Donoughue *et al* 2012), nuclear power (NETL 2012a, Warner and Heath 2012), CSP (Burkhardt *et al* 2012), PV (Hsu *et al* 2012, Kim *et al* 2012), and wind (Dolan and Heath 2012). We assume that, on first order and holding all other aspects of equation (1) constant, each performance parameter affects $e_{lifetime}$ or water$_{OP}$ (as physically appropriate) as:

$$e_{lifetime,h} = e_{lifetime,o} * \frac{p_h}{p_o}$$

(2)

Or

$$water_{OP,h} = water_{OP,o} * \frac{p_o}{p_h}$$

(3)

Where p is the relevant parameter, the subscript o signifies an original value, and the subscript h signifies a harmonized value. Because

$e_{lifetime}$ is increasing in each parameter p, the ratio $\left(\frac{P_h}{P_o}\right)$ scales $e_{lifetime,o}$ to the proportional change in parameter value. In contrast, generated electricity is in the implicit denominator of water$_{OP}$, so the inverse ratio

$\left(\frac{P_o}{P_h}\right)$ scales water$_{OP,o}$ to the proportional change in parameter value. The supplementary data (available at stacks.iop.org/ERL/8/015031/mmedia) presents all collected estimates as well as statistics on the availability of information for harmonization; the majority of estimates did not have relevant information and cannot be harmonized.

Table 1: Performance parameters used for harmonization and the life cycle stage in which each is applied

Parameter[a]	Value[b]	Stages for which parameter applies		
		Fuel cycle	Power plant	Operations
Thermal efficiency				
Coal: PC	35.4% (LHV), 34.3% (HHV)	x[c]	x[c]	x[c]
Coal: SC	39.9% (LHV), 38.4% (HHV)	x[c]	x[c]	x[c]
Coal: IGCC	39.8% (LHV), 38.5% (HHV)	x[c]	x[c]	x[c]
Coal: CFB	38.3% (LHV), 34.8% (HHV)	x[c]	x[c]	x[c]
Natural gas: CC	51.0% (HHV)	x	x	x
Natural gas: CT	33.0% (HHV)	x	x	x
Nuclear: fuel conversion	2.81 kg U_3O_8/kg $UF_{6(natural)}$	x		
Nuclear: fuel enrichment (diffusion)	10.4 kg $UF_{6(natural)}$/kg $UF_{6(enriched)}$	x		
Nuclear: fuel enrichment (centrifugal)	10.8 kg $UF_{6(natural)}$/kg $UF_{6(enriched)}$	x		
Nuclear: fuel fabrication	3.42 kg $UF_{6(enriched)}$/kg UO_2	x		
Nuclear: fuel use	0.004 33 kg UO_2 MWh^{-1}	x		
Fuel heat content				
Coal	21.01 MMBtu/ton (LHV)	x		
Natural gas	1031 Btu/scf (HHV)	x		
Solar-to-electric efficiency				
CSP: trough	15.0%		x	x
CSP: power tower	20.0%		x	x
PV: performance ratio	80%		x	x
PV: m-Si	13.0%		x	x
PV: p-Si	12.3%		x	x
PV: a-Si	6.3%		x	x
PV: CdTe	10.9%		x	x
PV: CIGS	11.5%		x	x
Solar resource				
CSP	2400 kWh m^{-2} yr^{-1}		x	x
PV	1700 kWh m^{-2} yr^{-1}		x	x

Capacity factor				
Coal	85%		×	
Natural gas	85%		×	
Nuclear	92%		×	
Wind: onshore	30%		×	
Wind: offshore	40%		×	
Power plant lifetime				
Coal	30 yr		×	
Geothermal	30 yr		×	
Natural gas	30 yr		×	
Nuclear	40 yr		×	
CSP	30 yr		×	
PV	30 yr		×	
Wind	20 yr		×	

[a]PC =pulverized coal, sub-critical; SC =pulverized coal, super-critical; CFB =circulated fluidized bed; IGCC =integrated gasification combined cycle; CC =combined cycle; CT =combustion turbine; m-Si =mono-crystalline silicon; p-Si =poly-crystalline silicon; a-Si =amorphous silicon; CdTe =cadmium telluride; CIGS =copper indium gallium selenide. [b]Parameters match those published in LCA harmonization and other benchmark studies on electricity generated by coal (MIT 2007, Whitaker et al 2012), natural gas (EIA 2011c, O'Donoughue et al 2012), nuclear power (NETL 2012a, Warner and Heath 2012), CSP (Burkhardt et al 2012), PV (Hsu et al 2012, Kim et al 2012), and wind (Dolan and Heath2012). We base the lifetime of geothermal on the lifetime used for other technologies. [c]Although LHV is the preferred measure for all life cycle stages of electricity generation by coal, data limitations require the use of HHV for the operations stage.

We develop estimates for each major life cycle stage for each generation technology (where coal, natural gas, and nuclear power have three major stages and all other generation technologies have two, as shown in figure 2). In many cases, such as different fuel extraction methods, cooling technologies, and generation prime movers, distinct production pathways have differentiated water use characteristics. We analyze such choices separately where data provide sufficient detail and aggregate such choices where data are more limited. For each production pathway option, we select the median estimate as reflecting the central tendency of the distribution of available data. Although gathered estimates are not a random sample of actual water use factors from the existing stock of generation facilities, we consider the median to be a reasonable representation across multiple references and technological variability within a category. Finally, we aggregate our selected, harmonized estimates into full life cycle water use estimates and investigate the sensitivity of life cycle water use to different values of the performance parameters shown in table 1.

The broad scope of this analysis necessitates important caveats and assumptions. Studies demonstrate considerable methodological inconsistency, and our attempts to address these through harmonization are limited by available information in the source literature. For example, the majority of thermoelectric operational water use estimates are not accompanied by thermal efficiency data so cannot be harmonized on this parameter. Reported boundaries around life cycle stages, including whether estimates include indirect water use, differ across references, but some references lack clear descriptions of the boundaries applied. As a result, we may overestimate life cycle water use in some cases where overlap leads to double counting and underestimate it in other cases where gaps in our estimate arise from processes being excluded from source data. Because available information suggests that most, but not all, reported estimates exclude indirect water use, we likely underestimate total attributable water use but overestimate on-site water use throughout the life cycle stages of electricity generation technologies. Although the magnitude of these errors is unknown, one expects it to correlate with an electricity generation technology's energy return on energy invested. In addition, although recycling and the use of degraded water both can dramatically reduce the amount of water used in multiple life cycle stages, particularly for withdrawals, we do not explicitly address these technological advances. Water use can vary substantially owing to site-specific differences such as local climate conditions, the age of equipment, and characteristics of the water source (Yang and Dziegielewski 2007) and the application of different environmental compliance technologies; the gross level of analysis presented here necessarily neglects such considerations. The assumption that performance parameters act proportionally upon water use factors does not account for non-linear effects; for example, higher irradiation increases CSP output directly but also may reduce the efficiency if operating temperatures are also raised (Turchi et a'2010). Finally, the estimates provided are neither predictions nor meant to exactly characterize all potential examples of deployment of a certain technology.

RESULTS: WATER USE ACROSS INDIVIDUAL LIFE CYCLE STAGES

Tables 2–12 present summary statistics of harmonized estimates of water consumption and withdrawal for major life cycle stages and production pathways for each generation technology, using the performance parameters shown in table 1. The full data these statistics summarize are available in the supplementary data (available at stacks.iop.org/ERL/8/015031/mmedia). Although median estimates are selected to represent each category, ranges reflect not only variability in the analytical reliability of collected estimates but also the aggregation of many potential sub-categories within technologies and life cycle stages. However, the minimum and maximum in the available literature may not represent the true minimum or maximum considering all deployment conditions, technological permutations, etc. Reflecting both the variability and uncertainty in the estimates for broad technology categories, reported results are limited to two significant digits.

Table 2: Summary statistics of selected, harmonized estimates of water consumption and withdrawal for major life cycle stages and production pathways for coal-fired electricity generation

	Sub-category[a]	Consumption (gal MWh^{-1})[b]				Withdrawal (gal MWh^{-1})[b]			
		Median	Min	Max	n^c	Median	Min	Max	n^c
Fuel cycle[d,e,f]	Surface mining	22	6	58	7	22	6	60	7
	Underground mining	56	17	230	7	57	17	230	7
Power plant[f]	Upstream and downstream[g]	1	<1[h]	25	8	1	<1[h]	12	8
Operations	PC: cooling tower	530	200	1300	20	660	460	1 200	21
	PC: open loop cooling	140	71	350	11	35 000	15 000	57 000	16
	PC: pond cooling	740	300	1000	11	10 000	300	26 000	10
	PC + CCS: cooling tower	940	900	940	3	1 300	1 200	1 400	3
	SC: cooling tower	500	460	590	7	600	580	670	7

	SC: open loop cooling	100	64	120	3	23 000	23 000	23 000	3
	SC: pond cooling	42	4	64	3	15 000	15 000	15 000	3
	SC + CCS: cooling tower	880	850	910	2	1 100	1 100	1 100	3
	CFB: cooling tower	560	560	560	1	1 000	1 000	1 000	1
	CFB: open loop cooling	210	210	210	1	20 000	20 000	20 000	1
	IGCC: cooling tower	320	35	440	14	390	160	6 700	16
	IGCC + CCS: cooling tower	550	520	600	4	640	480	740	7

[a]PC =pulverized coal, sub-critical; SC =pulverized coal, super-critical; CFB =circulated fluidized bed; IGCC =integrated gasification combined cycle; CCS =carbon capture and sequestration. [b]Statistics based on harmonized estimates, with respect to life cycle stage boundaries as well as relevant parameters shown in table 1. [c]For estimates constructed from multiple disaggregated stages or processes (reported in table 3), 'n' reports the average number of estimates over each of the stages, plus any included estimates that are not disaggregated. For categories with exactly 2 estimates, the median is defined as the arithmetic mean. [d]Fuel cycle estimates include estimates constructed from estimates for individual stages reported in table 3 in addition to estimates only for aggregated fuel cycle water use. [e]All fuel cycle estimates assume train transportation; mine-mouth conversion to electricity would decrease estimates negligibly and slurry pipeline transport would increase estimates substantially. [f]Fuel cycle and power plant estimates are harmonized to the thermal efficiency of a sub-critical pulverized coal power plant. [g]Power plant includes both upstream water use estimates (primarily for dust suppression during plant construction but also for manufacturing power plant raw materials) and downstream water use estimates (for decommissioning power plants). The latter contributes negligibly to the total for this life cycle stage. [h]<1 designates a value between 0.1 and 0.5 (due to rounding), and ≪1 designates a value less than 0.1.

Coal

Coal fuel cycle water factors, shown in table 2, are differentiated between surface and underground mining. Based on available data, we estimate that the fuel cycle uses approximately 22 gal MWh^{-1} with surface mining or 56 gal MWh^{-1} with underground mining, based primarily on US mining data and with estimates constructed from the individual process stages shown in table 3. Most water during extraction is used for dust suppression in mines and on roads, and higher surface mining estimates include water used for land reclamation whereas lower ones do not. We assume consumption equals withdrawals because consumption is often difficult to measure for mines; for example, NETL (2010a) reports that 'no specific data were located on the water consumed during mine operations...(consumption data) could not be separated from the storm water output' (p 33).

Table 3: Summary statistics of selected, harmonized estimates of water consumption and withdrawal for major production pathways in the coal fuel cycle

	Consumption (gal MWh^{-1})[a]				Withdrawal (gal MWh^{-1})[a]			
	Median	**Min**	**Max**	**n**	**Median**	**Min**	**Max**	**n**
Extraction (surface)[b]	3	<1[c]	13	9	3	<1[c]	13	9
Extraction (underground)[b]	27	8	180	8	27	8	180	8
Extraction (type not specified)[b]	45	12	120	4	45	12	120	4
Processing[b]	18	9	1000	8	18	9	1000	8
Transport (train)	<1[c]	≪1[c]	1	3	1	<1[c]	2	3
Transport (slurry)[b]	110	100	410	6	110	100	410	6

[a]Statistics based on harmonized estimates, with respect to life cycle stage boundaries as well as relevant parameters shown on table 1; estimates are harmonized to the thermal efficiency of a sub-critical pulverized coal power plant. This table does not include estimates that are reported only for the entire fuel cycle. [b]Reflecting data limitations and the nature of water use, we assume withdrawal and consumption are equal for all estimates in this category. [c]<1 designates a value between 0.1 and 0.5 (due to rounding), and ≪1 designates a value less than 0.1.

We omit older estimates of withdrawals up to 17 000 gal MWh[-1] for coal cleaning (Tolba 1985) in favor of newer, better documented estimates and therefore estimate that processing, which may or may not occur at the mine, contributes a median of 18 gal MWh[-1] to the fuel cycle total. In the final fuel cycle stage of coal transportation, although slurry pipelines consume 110 gal MWh[-1] of water, train transport is more common currently, with the median water use reported in table 2 corresponding to a 205 mile transport distance (NETL 2010c).

When amortized to the power plant's lifetime generation, upstream and downstream water use for the coal power plant's equipment is negligible. In contrast, coal power plant cooling requires hundreds to thousands of gallons withdrawn and consumed per MWh. Water use factors vary substantially by cooling technology, with open loop cooling (also known as once-through cooling) requiring much greater withdrawals and recirculating cooling towers consuming relatively more water, as shown in table 2 and described in more detail elsewhere (Macknick *et al* 2012). Estimates for pond-cooled systems vary widely, because they can be operated similarly to either once-through or recirculating tower systems.

Generally, more efficient combustion technologies (e.g., integrated gasification combined cycle (IGCC)) require less water per unit generation for cooling than less efficient ones (e.g., sub-critical pulverized coal). Variation in reported estimates arises from variation in specific operating conditions as well as whether estimates include non-cooling water uses. For example, additional operational needs, such as pollution controls specific to coal, can be substantial; one study reports 155 gal MWh[-1] for coal–ash handling, 12 gal MWh[-1] for NO_x control, and 60 gal MWh[-1] for SO_x scrubbing (TWDB 2003). Flue gas desulfurization increases water use by about 40 gal MWh[-1] using dry technology and about 70 gal MWh[-1] using wet technology (NETL 2009a).

As shown in table 4, collected data suggest that carbon capture and sequestration (CCS) can increase operations water consumption by about 75% and water withdrawal by between 64% and 97%, due to a combination of lower efficiencies and additional process demands for certain CCS technologies. Efficiency penalties also increase the fuel cycle and power plant equipment life cycle water use per generated MWh. However, the water use of all CCS technologies has not been characterized in the literature, and technologies with different efficiency and operational characteristics would lead to different relative water impacts.

Table 4: Estimated effect of carbon capture and sequestration (CCS) on life cycle water use for coal- and natural gas-fired electricity generation

		Pulverized coal (sub-critical)	Pulver-ized coal (super-critical)	Circulat-ing fluid-ized bed	Integrated gasifi-cation combined cycle	Natural gas combined cycle
Thermal efficiency (HHV)[a]	w/o CCS	34.3%	38.5%	34.8%	38.4%	51.0%
	w/ CCS	25.1%	29.3%	25.5%	31.2%	43.9%
	Change	−27%	−24%	−27%	−19%	−14%
Power plant, consumption (gal MWh^{-1})[b]	w/o CCS	0.9	0.8	0.8	0.8	0.4
	w/ CCS	1.2	1.0	1.1	1.0	0.8
	Change	37%	31%	36%	23%	100%
Power plant, withdrawal (gal MWh^{-1})[b]	w/o CCS	1.3	1.2	1.2	1.2	0.6
	w/ CCS	1.8	1.5	1.6	1.5	1.0
	Change	37%	31%	36%	23%	67%
Fuel cycle, consumption (gal MWh^{-1})[c]	w/o CCS	22	19	20	20	4
	w/ CCS	30	25	27	24	5
	Change	37%	31%	36%	23%	16%
Fuel cycle, withdrawal (gal MWh^{-1})[c]	w/o CCS	22	19	20	20	5
	w/ CCS	31	25	27	24	6
	Change	37%	31%	36%	23%	16%
Operations, consumption (gal MWh^{-1})[d]	w/o CCS	530	500	560	320	210
	w/ CCS	940	880	980[f]	550	380
	Change	77%	76%	75%[f]	72%	81%
	n[e]	3	2	0[f]	4	2
Operations, withdrawal (gal MWh^{-1})[d]	w/o CCS	660	600	1000	390	250

w/ CCS	1300	1100	1800[f]	640	510
Change	97%	83%	81%[f]	64%	104%
n^e	3	3	0[f]	7	3

[a]As reported in MIT (2007) for coal technologies and NETL (2010b) for natural gas technology. [b]We estimate power plant water use for different coal-fired generation technologies with and without CCS by adjusting the parameter for thermal efficiency from our base case (for sub-critical pulverized coal without CCS) to the reported thermal efficiencies. For natural gas-fired generation with CCS, we use the estimate reported by NETL (2010b). [c]We estimate fuel cycle water use for different generation technologies with and without CCS by adjusting the parameter for thermal efficiency from our base cases for coal- and natural gas-fired generation to the reported thermal efficiencies. [d]Estimated using median reported values as reported in tables 2 and 5. [e]'n' reports the number of operations water use estimates for each prime mover technology with CCS, as reported in tables 2 and 5. [f]We estimate water use during operations of a circulated fluidized bed power plant equipped with CCS based on the weighted average of the other generation technology's estimates, due to a lack of reported estimates.

Natural Gas

The natural gas-fired electricity life cycle also includes a fuel cycle, with the key water-relevant fuel cycle distinction being the use of hydraulic fracturing for shale gas extraction versus other conventional methods of extracting gas sources, as shown in tables 5 and 6. For hydraulic fracturing in shale, reported water use ranges from 300 000 gallons per well (Noble Energy Inc. and CSU2012) to nearly 9 million gallons per well (TWDB 2012), with 50% of collected data reporting between 2 and 5 million gallons per well. Amortized by play-specific estimated ultimate recovery (EUR) of a well as reported in EIA (2011c), this corresponds to a median of 16 gal MWh^{-1} for shale gas hydraulic fracturing. Reflecting recent interest on the topic (e.g., GAO 2012, JISEA 2012, TWDB 2012), we found many estimates for water use in shale gas hydraulic fracturing and display these in figure 3. The variation in estimates corresponds to the range of water volumes associated with technological differences; to variations in EUR, which can vary by as much as a factor of 10 across wells even within the same formation (GAO 2012); and to variation in other factors such as well length.

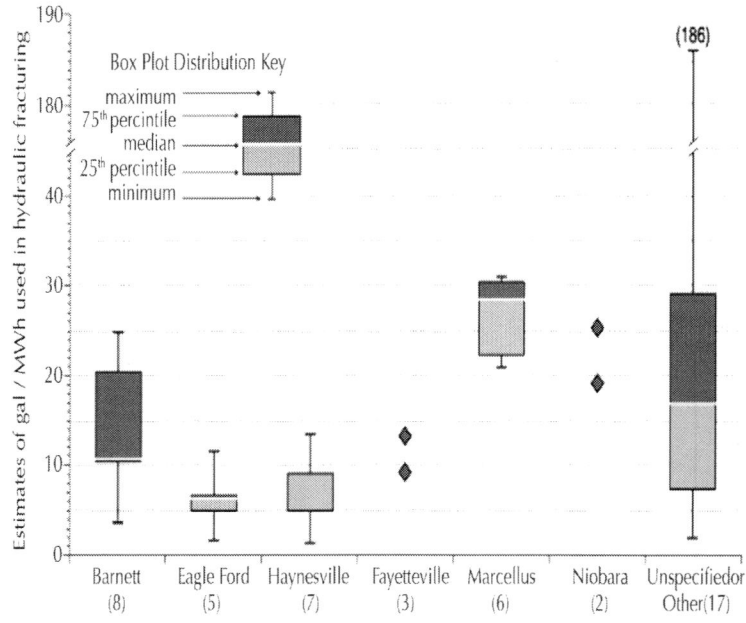

Figure 3: Distributions of estimates of water use for hydraulic fracture stimulation for shale gas extraction, expressed as a function of generated electricity, demonstrate variability both within and across different shale plays. Estimates do not include water used for drilling. The number of estimates for each play is presented in parentheses. The broken y-axis accommodates one outlier estimate reported as the upper limit of water use by IEA (2012). This figure is based on author analysis of collected estimates from 17 references; all fracturing (shale) estimates provided in the Natural Gas–Fuel Cycle–consumption table in the supplemental data (available at stacks.iop.org/ERL/8/015031/mmedia) are represented in this figure.

Table 5: Summary statistics of selected, harmonized estimates of water consumption and withdrawal for major life cycle stages and production pathways for natural gas-fired electricity generation

	Sub-category[a]	Consumption (gal MWh^{-1})[b]				Withdrawal (gal MWh^{-1})[b]			
		Median	Min	Max	n[c]	Median	Min	Max	n[c]
Fuel cycle[d,e]	Conventional natural gas[f]	4	1	26	9	5	4	34	8
	Shale gas	16	3	210	20	17	5	220	18
Power plant[e]	Upstream and downstream[g]	1	<1[h]	1	2	<1[h]	<1[h]	1	3
Operations	CC: cooling tower	210	47	300	19	250	150	760	16
	CC: dry cooling	4	4	120	4	4	≪1[h]	4	2
	CC: open loop cooling	100	20	230	8	9 000	7 200	21 000	7
	CC: pond cooling	240	240	240	2	6 000	6 000	6 000	2
	CC + CCS: cooling tower	380	380	380	2	510	490	510	3
	CT	50	50	340	3	430	430	430	1
	Steam: cooling tower	730	560	1100	8	1 200	1 200	1 200	2
	Steam: open loop cooling	290	190	410	6	36 000	35 000	37 000	2
	Steam: pond cooling[i]	270	270	270	1	270	270	270	1

[a]CC =combined cycle; CT =combustion turbine; CCS =carbon capture and sequestration. [b]Statistics based on harmonized estimates, with respect to life cycle stage boundaries as well as relevant parameters shown in table 1. [c]For estimates constructed from multiple disaggregated stages or processes (reported in table 6), 'n' reports the average number of estimates over each of the stages. For categories with exactly 2 estimates, the median is defined as the arithmetic mean. [d]Fuel cycle estimates consists of estimates constructed from estimates for the individual stages within the fuel cycle (reported in table 6). All fuel cycle estimates assume pipeline transportation; estimates for water use in the fuel cycle of liquid natural gas (LNG) range widely and could potentially increase estimates substantially. [e]Fuel cycle and power plant

estimates are harmonized to the thermal efficiency of a combined cycle plant. [f]We define conventional natural gas as that not requiring fracture stimulation.[g]Power plant includes both upstream water use estimates (primarily for dust suppression during construction but also for the water use for manufacturing power plant raw materials) and downstream water use estimates (for water used in decommissioning power plants). The latter contributes negligibly to the total for this life cycle stage. [h]<1 designates a value between 0.1 and 0.5 (due to rounding), and ≪1 designates a value less than 0.1. [i]Reflecting data limitations and the nature of water use, we assume withdrawal and consumption are equal for all estimates in this category.

Table 6: Summary statistics of selected, harmonized estimates of water consumption and withdrawal for major production pathways in the natural gas fuel cycle

	Consumption (gal MWh^{-1})[a]				Withdrawal (gal MWh^{-1})[a]			
	Median	Min	Max	n[b]	Median	Min	Max	n
Drilling[c]	1	≪1[e]	19	29	1	≪1[e]	19	29
Fracturing (other fracture stimulated gas)[c,d]	<1[e]	≪1[e]	2	4	<1[e]	≪1[e]	2	4
Fracturing (shale gas)[c]	12	1	186	49	12	1	186	49
Processing[c]	<1[e]	<1[e]	<1[e]	1	<1[e]	<1[e]	<1[e]	1
Transport (pipeline)	3	1	6	2	4	4	13	3
Transport (liquefied natural gas)	1	1	1	1	8	8	8	1

[a]Statistics based on harmonized estimates, with respect to life cycle stage boundaries as well as relevant parameters shown in table 1; estimates are harmonized to the thermal efficiency of a combined cycle natural gas power plant. [b]For categories with exactly 2 estimates, the median is defined as the arithmetic mean. [c]Reflecting data limitations and the nature of water use, we assume withdrawal and consumption are equal for all estimates in this category. [d]The 'other fracture stimulated gas' category includes two estimates for tight gas and two reported for 'conventional gas with fracture stimulation' (all from IEA (2012)). [e]<1 designates a value between 0.1 and 0.5 (due to rounding), and ≪1 designates a value less than 0.1.

After extraction, natural gas is processed to bring it to pipeline quality. Although three older references (DOE 1983, Tolba 1985, Gleick 1994) agree upon a relatively high water usage of 11 gal MWh^{-1} for this processing, we defer to the more recent NETL (2010d) assessment that processing requires an equivalent of only 0.1 gal MWh^{-1}. Although this estimate is only for natural gas sweetening, it is unclear that any other stages of the natural gas processing stage use water; note that Grubert *et al* (2012) estimate no water use for natural gas processing. In all, our analysis suggests 4 gal MWh^{-1} consumed and 5 gal MWh^{-1}withdrawn in the fuel cycle of conventional natural gas, and 16 gal MWh^{-1} and 17 gal MWh^{-1} in that of shale gas.

As with most other thermoelectric technologies, we estimate water use in the power plant equipment's life cycle as negligible (1 gal MWh^{-1} or less), and operational water use far exceeds that for other life cycle stages in most cases, with important differences among cooling technologies. Reflecting the high thermal efficiencies of combined cycle natural gas plants relative to coal combustion, water used in natural gas operations is approximately one-half to one-third that for coal for a given cooling technology. Less efficient gas combustion technologies have higher operational water use estimates, and as shown in table 4, CCS technology can increase operational water use by as much as a factor of two.

Nuclear

The uranium fuel cycle includes extraction, numerous processing steps, and greater end-of-life considerations than other fuels. As shown in table 7, our analysis suggests that the nuclear power fuel cycle typically withdraws 56 gal MWh^{-1} water with centrifugal enrichment and 140 gal MWh^{-1} (87 gal MWh^{-1} of which is consumed) for gaseous diffusion enrichment. Due to data limitations, including consumption estimates that often exceed estimates reported for withdrawal, we assume that withdrawal equals consumption for all fuel cycle stages except gaseous diffusion enrichment. Consistent with other technologies, we consider only externally sourced water; estimates including produced water for *in situ* leaching exceed those shown here by approximately 70 times (e.g., Mudd and Diesendorf 2009). We do not distinguish between extraction methods; as table 8 shows, external water use estimates are not significantly differentiated by extraction method. End-of-life water use for nuclear power is uncertain; currently implemented fuel management options are at the low end of estimates (e.g., 1 gal MWh^{-1} withdrawal (Schneider *et al* 2010)), whereas potential fuel recycling is estimated at 720 gal MWh^{-1} withdrawal (NETL 2012a).

Table 7: Summary statistics of selected, harmonized estimates of water consumption and withdrawal for major life cycle stages and production pathways for nuclear power

	Sub-category	Consumption (gal MWh^{-1})[a]				Withdrawal (gal MWh^{-1})[a]			
		Median	Min	Max	n[b]	Median	Min	Max	n[b]
Fuel cycle[c]	Centrifugal enrichment	56	13	300	5	56	13	300	5
	Diffusion enrichment	87	42	330	5	140	62	410	5
Power plant	Upstream and downstream[d]	<1[e]	<1[e]	<1[e]	2	<1[e]	<1[e]	<1[e]	2
Operations	Cooling tower	720	580	890	9	1 100	800	2 600	7
	Open loop cooling	400	100	400	5	47 000	23 000	60 000	12
	Pond cooling	610	400	720	4	1 100	500	13 000	4

[a]Statistics based on harmonized estimates, with respect to life cycle stage boundaries as well as relevant parameters shown in table 1. [b]For estimates constructed from multiple disaggregated stages or processes (reported in table 8), 'n' reports the average number of estimates over each of the stages. For categories with exactly 2 estimates, the median is defined as the arithmetic mean. [c]Fuel cycle estimates consists of estimates constructed from the individual stage estimates. All fuel cycle estimates represent a combined estimate for different extraction types and storage and disposal for spent fuel; reprocessing and recycling of spent fuel is not currently practiced in the United States but would increase estimates substantially. [d]Due to limited information, upstream water use includes estimates only for manufacturing power plant raw materials. In addition, although the stage is expected to be negligible, no estimates for downstream water use were found. [e]<1 designates a value between 0.1 and 0.5 (due to rounding), and ≪1 designates a value less than 0.1.

Table 8: Summary statistics of selected, harmonized estimates of water consumption and withdrawal for major production pathways in the nuclear fuel cycle

	Consumption (gal MWh^{-1})[a]				Withdrawal (gal MWh^{-1})[a]			
	Median	Min	Max	n[b]	Median	Min	Max	n[b]
Extraction (*in situ* leaching)[c,d]	18	13	23	2	18	13	23	2
Extraction (surface)[c]	32	4	92	6	32	5	92	6
Extraction (underground)[c]	30	<1[g]	240	4	30	<1[g]	240	4
Extraction (type not specified)[c]	15	15	15	1	15	15	15	1
Processing (milling)[c]	11	3	29	6	11	3	29	6
Processing (conversion)[c]	10	4	13	3	10	4	13	3
Processing (centrifugal enrichment)[c]	4	3	6	3	4	3	6	3
Processing (diffusion enrichment)	35	32	37	2	83	51	120	2
Processing (fuel fabrication)[c]	1	1	3	4	1	1	3	4
End-of-life (storage and disposal)[c,e]	3	1	5	3	3	1	5	3
End-of-life (reprocessing spent fuel)[f]	7	7	7	1	720	720	720	1

[a]Statistics based on harmonized estimates, with respect to life cycle stage boundaries as well as relevant parameters shown in table 1. [b]For categories with exactly 2 estimates, the median is defined as the arithmetic mean. [c]Reflecting data limitations and the nature of water use, we assume withdrawal and consumption are equal for all estimates in this category. [d]For *in situ* leaching, only external water use is considered. The inclusion of produced water can lead to estimates on the order of 70 times higher. [e]Storage and disposal includes estimates both of 'temporary' storage on site and also the hypothetical Yucca Mountain storage facility. [f]Reprocessing of fuel is based on a hypothetical facility (NETL 2012a).[g]<1 designates a value between 0.1 and 0.5 (due to rounding), and ≪1 designates a value less than 0.1.

The large lifetime output of nuclear power plants leads to negligible estimates for the power plant equipment life cycle. Estimated operational water requirements, however, are around one or more orders of magnitude higher than fuel cycle estimates. As for other generation technologies, we estimate that the proportion of withdrawn water that is consumed is higher for cooling towers than it is for open loop cooling, and the proportion of consumption for pond cooling is in between the two.

Concentrating Solar Power (CSP)

CSP has no fuel cycle, but CSP power plant life cycle estimates (in gal MWh^{-1}) are higher than for the non-renewable thermoelectric technologies. This likely reflects the lower lifetime output over which upstream use is amortized and the use of specialty chemicals requiring more water than commodities used in typical thermoelectric technologies. Although Inhaber (2004) estimates about 1 gal MWh^{-1} for the power plant equipment life cycle based on material volumes, other references estimate water consumption just for construction between 1 and 80 gal MWh^{-1}. We omit these estimates and instead base our median water use estimate of 160 gal MWh^{-1} only on references reporting more comprehensively for the power plant life cycle (Burkhardt et al 2011, NETL 2012b). As a thermoelectric generation technology, CSP withdraws similar amounts of operational water to pulverized coal technology, with important differences by cooling technology. However, because CSP systems commonly are located in remote areas and use evaporation ponds for water disposal, consumption volumes typically equal withdrawals regardless of technology (Solar Millennium LLC 2008).

Table 9: Summary statistics of selected, harmonized estimates of water consumption and withdrawal for major life cycle stages and production pathways for CSP-generated electricity

	Sub-category	Consumption (gal MWh^{-1})[a]				Withdrawal (gal MWh^{-1})[a]			
		Median	Min	Max	n[b]	Median	Min	Max	n[b]
Power plant	Upstream and downstream[c]	160	80	170	3	160	99	170	3
Operations	Dish stirling[d]	5	5	5	2	5	5	5	2

Fresnel[d]	1000	1000	1000	1	1000	1000	1000	1
Power tower: cooling tower	810	740	860	5	740	740	740	1
Power tower: dry cooling[d]	26	26	26	1	26	26	26	1
Power tower: hybrid cooling[d]	170	90	250	2	170	90	250	2
Trough: cooling tower	890	560	1900	26	960	870	1100	2
Trough: dry cooling	78	32	140	20	78	33	79	11
Trough: hybrid cooling	340	110	350	3	340	340	340	1

[a]Statistics based on harmonized estimates, with respect to life cycle stage boundaries as well as relevant parameters shown in table 1. [b]For estimates constructed from multiple disaggregated stages or processes, 'n' reports the average number of estimates over each of the stages. For categories with exactly 2 estimates, the median is defined as the arithmetic mean. [c]Power plant includes both upstream water use estimates (using only those that include manufacturing in addition to construction) and downstream water use estimates (for water used in dismantling and disposal of power plants). The latter contributes negligibly to the total for this life cycle stage. Estimates are harmonized to the solar-to-electric efficiency of a trough power plant. We include consumption estimates from Burkhardt et al (2011) in the withdrawal category due to lack of available data on withdrawals; therefore, withdrawals might be underestimated. [d]Reflecting data limitations and the nature of water use, we assume withdrawal and consumption are equal for all estimates in this category.

Geothermal

Usable data on geothermal power's water use are limited, as shown in table 10. From available data, we estimate that 3 gal MWh^{-1} of water are withdrawn and 2 gal MWh^{-1} are consumed in the power plant equipment life cycle, with water use dominated by drilling and cementation in plant construction. As described in more detail elsewhere (Clark et al 2011, Macknick et al 2012), operational water use varies by more than an order of magnitude corresponding both to technology configurations (e.g., dry steam, binary, and flash) and to local contexts. Consistent with Macknick et al (2012) and our treatment of other technologies, we screened out many published estimates of geothermal operations water use that included geothermal fluids in operational water requirements and focus only on estimates of external water required for operations. Estimates of water use that include geothermal fluids in operational water requirements report water consumption values between roughly 2000 gal MWh^{-1} to 4000 gal MWh^{-1}, with estimates of EGS even higher (Macknick *et al* 2011). In addition, because the limited data led to results conflicting with physical laws, we include an estimate for a tower-cooled binary power plant that normally would be excluded by our screens because the value (700 gal MWh^{-1}) is estimated from a graphic in the original source (Kozubal and Kustcher 2003).

Table 10: Summary statistics of selected, harmonized estimates of water consumption and withdrawal for major life cycle stages and production pathways for geothermal power-generated electricity

	Sub-category[a]	Consumption (gal MWh^{-1})[b]				Withdrawal (gal MWh^{-1})[b]			
		Median	Min	Max	n[c]	Median	Min	Max	n[c]
Power plant	Upstream and downstream[d]	2	2	2	1	3	<1[g]	10	11
Operations[e]	Binary: hybrid cooling[f]	460	220	700	2	460	220	700	2
	Binary: dry cooling[f]	290	270	630	3	290	270	630	3
	Flash	11	5	360	5	18	11	25	2
	EGS: dry cooling[f]	510	290	720	2	510	290	720	2

[a]EGS =enhanced geothermal system. [b]Statistics based on harmonized estimates, with respect to life cycle stage boundaries as well as relevant parameters shown in table 1. [c]For categories with exactly 2 estimates, the median is defined as the arithmetic mean. [d]Due to limited information, upstream water use includes estimates only for manufacturing power plant raw materials and construction. In addition, although the stage is expected to be negligible, no estimates for power plant downstream water use were found. [e]We omit many collected estimates, which conflate produced water with external water; only external water use is reported in the table. [f]Reflecting data limitations and the nature of water use, we assume withdrawal and consumption are equal for all estimates in this category. [g]<1 designates a value between 0.1 and 0.5 (due to rounding), and ≪1 designates a value less than 0.1.

Photovoltaics (PV)

As shown in table 11, estimates of water withdrawal for PV's power plant equipment life cycle vary widely, from 1 to 1600 gal MWh^{-1}, with the majority near the median estimate of 94 gal MWh^{-1} for crystalline silicone (C-Si). Of the few consumption estimates available, those matched to withdrawal estimates suggest approximately 30% of withdrawn water is consumed (Genesee County Economic Development Center 2011). Because the water use for the PV power plant potentially dominates that for other PV life cycle stages, this variation argues for further study; processing silicon into PV equipment involves numerous stages with rapidly developing techniques that are much less established than manufacturing processes used for other generation technologies, such as the fabrication of steel components. Given this uncertainty, we combine a variety of individual PV technologies, mostly thin-films, into a single 'other' category.

Table 11: Summary statistics of selected, harmonized estimates of water consumption and withdrawal for major life cycle stages and production pathways for PV-generated electricity

	Sub-category	Consumption (gal MWh^{-1})[a]				Withdrawal (gal MWh^{-1})[a]			
		Median	Min	Max	n[b]	Median	Min	Max	n
Power plant[c]	C-Si (crystalline silicone)	81	10	210	3	94	1	1600	24
	Other (primarily thin-film)	6	5	7	2	18	<1[d]	1400	19
Operations	Flat paneld[e]	6	1	26	9	6	1	26	9
	Concentrated PV[e]	30	24	78	4	30	24	78	4

[a]Statistics based on harmonized estimates, with respect to life cycle stage boundaries as well as relevant parameters shown in table 1. [b]For categories with exactly 2 estimates, the median is defined as the arithmetic mean. [c]Power plant estimates include both upstream (i.e., raw materials, manufacturing, construction, and transportation) and downstream (decommissioning) water use. Due to a lack of data, we assume for downstream processes that consumption is negligible and withdrawal is equivalent across sub-categories. [d]<1 designates a value between 0.1 and 0.5 (due to rounding), and ≪1 designates a value less than 0.1. [e]Reflecting data limitations and the nature of water use, we assume withdrawal and consumption are equal for all estimates in this category.

Water use for operations is minimal. Experimental evidence demonstrates that although frequent washing increases output, it likely leads to economic losses (Sahm *et al* 2005). DOE (2012) reports that few operators wash PV panels in actual practice. The higher water use of concentrated PV likely reflects certain shared operational characteristics with CSP, such as a need for mirror washing.

Wind

Although wind data come from few references, these estimates appear of high quality due to thorough information, including paired reporting of withdrawal and consumption, relatively close agreement on withdrawals across sources, and relatively common reporting for power plant downstream water use. We omit older estimates of manufacturing water use that are based on standard manufacturing practices and the bulk volume of materials (Inhaber 2004) in favor of detailed data from manufacturers and a national laboratory, resulting in a median withdrawal of 26 gal MWh^{-1} and consumption of 1 gal MWh^{-1}. Wind turbines require no fuel and little, if any, washing and maintenance, so operational water use is very low.

Table 12: Summary statistics of selected, harmonized estimates of water consumption and withdrawal for major life cycle stages and production pathways for wind-generated electricity

	Sub-category	Consumption (gal MWh^{-1})[a]				Withdrawal (gal MWh^{-1})[a]			
		Median	Min	Max	n[b]	Median	Min	Max	n[b]
Power plant	Upstream and downstream[c]	1	$\lll 1$[d]	9	12	26	13	83	19
Operations	Onshore	<1[d]	$\lll 1$[d]	2	10	1	1	1	2
	Offshore	$\lll 1$[d]	$\lll 1$[d]	1	4	2	$\lll 1$[d]	3	9

[a]Statistics based on harmonized estimates, with respect to life cycle stage boundaries as well as relevant parameters shown in table 1. [b]For estimates constructed from multiple disaggregated stages or processes, 'n' reports the average number of estimates over each of the stages. For categories with exactly 2 estimates, the median is defined as the arithmetic mean. [c]Power plant includes both upstream water use estimates (pertaining to manufacturing, materials, and construction) and downstream water use estimates (for water used in dismantling and disposal of power plants). The latter contributes negligibly to the total for this life cycle stage. [d]<1 designates a value between 0.1 and 0.5 (due to rounding), and $\lll 1$ designates a value less than 0.1.

RESULTS: WATER USE ACROSS THE FULL LIFE CYCLE

Figures 4 and 5 depict estimated total life cycle water consumption and water withdrawal, respectively, for selected production pathways for each generation technology, leveraging the best available evidence collected and screened herein. We construct life cycle water use by summing water use factors for relevant stages presented above using the consistent performance parameters presented in table 1 and using consistent definitions for each life cycle stage. These estimates are based on median values and thus ignore the important variation within estimates for each stage. More generally, life cycle water use estimates are a limited indicator of aggregate impact on water resources, given the critical spatial and temporal characteristics of resource demands and availability.

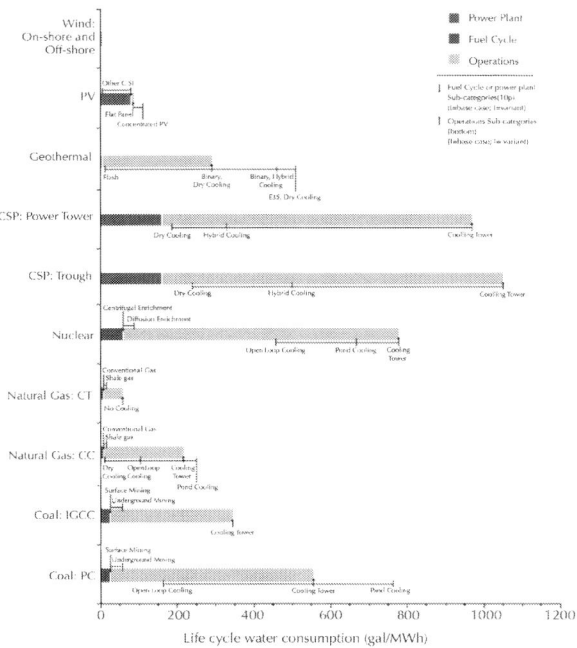

Figure 4: Estimated life cycle water consumption factors for selected electricity generation technologies, based on median harmonized estimates, demon-

strate significant variability with respect to technology choices. Base case estimates for each life cycle stage, presented in bold font, are held constant for estimating life cycle water consumption factors for other life cycle stages. Estimates for production pathway variants in fuel cycle or power plant (labeled on top of the bars) or operations (bottom) are labeled at points connected to the base case estimate with horizontal lines. Note: PV=photovoltaics; C-Si =crystalline silicone; EGS =enhanced geothermal system; CSP =concentrating solar power; CT =combustion turbine; CC =combined cycle; IGCC =integrated gasification combined cycle; and PC = pulverized coal, subcritical.

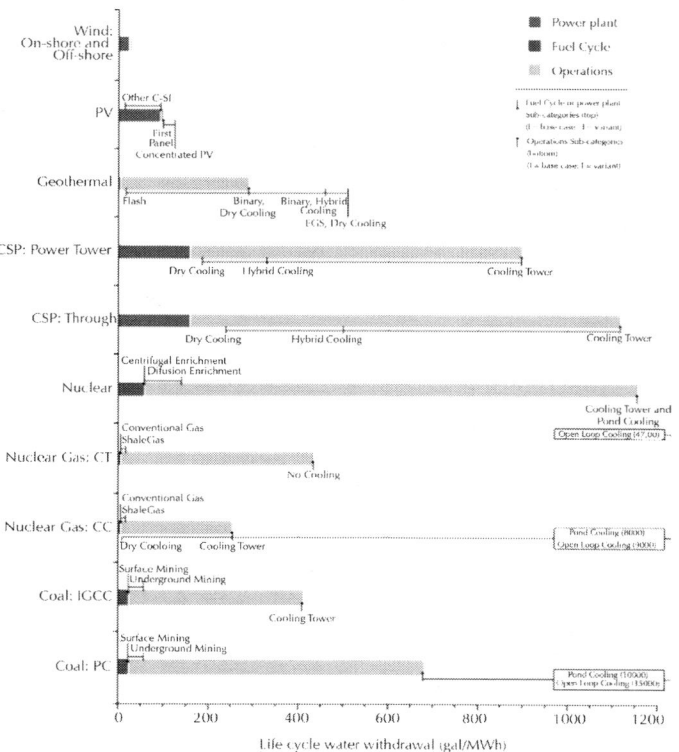

Figure 5: Estimated life cycle water withdrawal factors for selected electricity generation technologies, based on median harmonized estimates, demonstrate significant variability with respect to technology choices. Base case estimates for each life cycle stage, presented in bold font, are held constant for estimating life cycle water consumption factors for other life cycle stages. Estimates for production pathway variants in fuel cycle or power plant (labeled

on top of the bars) or operations (bottom) are labeled at points connected to the base case estimate with horizontal lines. Note: PV=photovoltaics; C-Si=crystalline silicone; EGS =enhanced geothermal system; CSP =concentrating solar power; CT =combustion turbine; CC =combined cycle; IGCC =integrated gasification combined cycle; and PC = pulverized coal, sub-critical.

Operations dominate the life cycle water use for most electricity production pathways, with the exceptions of dry-cooled thermoelectric technologies, PV, and wind. Accordingly, relative rankings of life cycle water use mirror those for the operations stage presented in Macknick *et al* (2012). For coal, natural gas, and nuclear, the fuel cycle contributes a small but non-negligible amount to total life cycle water use. For these technologies, power plant equipment life cycle water demands are negligible in relation to the life cycle total. In contrast, the power plant contributes a large portion of the total water use for the thermoelectric renewable technology of CSP, and represent the majority of life cycle water use for non-thermoelectric renewables (PV and wind). With the exception of prominent distinctions between withdrawal and consumption requirements for different cooling technologies, most estimates of water consumption and withdrawal across the life cycle of a given production pathway follow similar relative patterns to each other. In both figures 4 and 5, the relative rankings of water use across major generation technology categories switch according to production pathways.

Figure 6 demonstrates the sensitivity of the life cycle water consumption estimates shown in figure 4 to the selection of performance parameters for harmonization. The range of parameters used match the extremes found in the published literature or reported in reviews on electricity generated by coal (Whitaker *et al* 2012), natural gas (O'Donoughue *et al* 2012), nuclear power (NETL 2012a, Warner and Heath 2012), CSP (Burkhardt *et al* 2012, DOE 2012), PV (Hsu *et al* 2012), and wind (Dolan and Heath2012). We base the range for geothermal on the ranges used for other technologies.

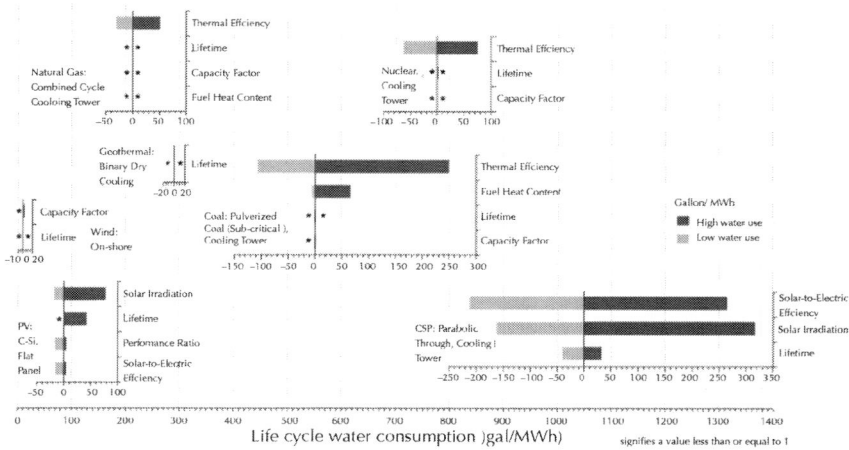

Figure 6: Sensitivity analysis reveals differing influence of certain performance parameters on the median estimate of life cycle water consumption factors for selected electricity generation technology production pathways. The figure depicts the variation of median life cycle stage estimates across a range of reasonable performance parameters. See text and section A.4 of the supplemental data (available at stacks.iop.org/ERL/8/015031/mmedia) for further details. Note: PV =photovoltaics; C-Si =crystalline silicone; and CSP = concentrating solar power.

Parameter values can alter the relative rankings of water consumption across major generation technology categories (e.g., coal versus nuclear). The relative sensitivity of the life cycle total consumption to performance parameter values corresponds to the relative contribution of the major life cycle stages to which they pertain. In addition to demonstrating a source of variation in published estimates, this figure reflects how operations characteristics interact with relative water use in other life cycle stages to influence water use per unit of electricity output. For example, variation in thermal efficiency corresponds with substantial variation in life cycle water use because the factor affects the amount of cooling water required in operations, but also how much fuel is required and therefore how much water is required for fuel extraction (which is measured in terms of water use per unit of fuel). Furthermore, factors affecting only the amortization of power plant equipment, such as lifetime or capacity factor, have little influence on life cycle water factors when the power plant equipment contributes only negligibly to the life cycle total.

CONCLUSIONS

In a water-constrained world, it is critical to understand how water is used throughout the entire life cycle of electricity generation. From a wide array of sources, we gathered available evidence for water use in any stage of the electricity generation life cycle for selected technologies. We screened and harmonized estimates to common performance parameters and boundaries and consolidated them into water consumption and withdrawal factors for major life cycle stages for each electricity generation technology considered.

This concise presentation of life cycle water use factors, built from a thorough review of the available literature, offers a unique and comprehensive look at the water requirements of different electricity generation and fuel supply choices. In many cases, operations dominate life cycle water use in absolute magnitude. However, the water implications of choices about electricity generation technologies clearly do not end at those due to the cooling water demands for thermoelectric generation. The fuel cycles of coal, natural gas, and nuclear power all require significant water volumes, and renewable generation technologies require significant water for manufacturing and construction. Such considerations may be important to the development both of local fuel resources and of local electricity generation capacity in regions with current or potential constraints on water resources. In general, based on review and harmonization of the available evidence, total life cycle water use across the generation technologies considered here appears lowest for electricity generated by photovoltaics and wind, and highest for nuclear and conventional coal technologies. Depending on cooling and prime mover technologies, natural gas and CSP technologies can be ranked either among the highest or lowest water users. For a given generation technology and cooling type, evaluated CCS technologies can increase operational water use by a factor of two and upstream water use by an amount proportional to the associated loss in efficiency.

Despite extensive collection, screening, and harmonization efforts, gathered estimates for most generation technologies and life cycle stages remain few in number, wide in range, and many are of questionable original quality. These constraints should be considered for proper interpretation and use of the results reported here in future analyses. For example, reflecting a general lack of rigorously tracked and recorded comparisons of consumption to withdrawal, median estimates of consumption in some cases exceed the corresponding median estimate for withdrawal and were thus adjusted in the results reported herein to conform to physical laws. Estimates for nearly all processes and life cycle stages vary significantly, reflecting a combination of issues including methodological inconsistency, sub-category heterogeneity, and the effect of local conditions on water use. Data limitations highlight the need for new sources of primary data for many life cycle stages of many generation technologies. Although most categories would benefit from new sources of more recent, well-documented primary data, the limitations suggest particular value from further research into areas with relatively large variation in estimates, such as the PV power plant equipment life cycle, or with few available estimates, such as the nuclear fuel cycle or the full life cycle of geothermal electricity generation.

This analysis establishes a foundation for estimating water requirements of different electricity generation choices. Estimation of water use for actual projects should use the most specific data possible, in light of the finding that the ranking of water use across generation technologies is not fixed but varies with production pathway and by specific performance parameters. However, this paper provides insight by consolidating and screening the wide breadth of available information into robust first order estimates of water used by archetypal production pathways across the life cycle. Improved understanding of water use can inform management of risks associated with water resource variability, within each part of the production pathway.

ACKNOWLEDGMENTS

This work was supported by the US Department of Energy under Contract No. DE-AC36-08-GO28308 with the National Renewable Energy Laboratory. We wish to thank Laura Vimmerstedt and Dan Bilello, whose comments helped to improve the manuscript. We also acknowledge the LCA Harmonization project team that developed the database of LCA publications (www.nrel.gov/harmonization), Alfred Hicks for polishing the graphics, and Judy Oberg for research assistance.

REFERENCES

1.   Averyt K, Macknick J, Rogers J, Madden N, Fisher J, Meldrum J and Newmark R 2013 Water use for electricity in the united states: an analysis of reported and calculated water use information for 2008 *Environ. Res. Lett.* 8 015001.

2.   Berndes G 2008 Future biomass energy supply: the consumptive water use perspective *Int. J. Water Resources Dev.* 24235–45

3.   Burkhardt J J, Heath G and Cohen E 2012 Life cycle greenhouse gas emissions of trough and tower concentrating solar power electricity generation: systematic review and harmonization *J. Ind. Ecol.* 16 S93–S109

4.   Burkhardt J J, Heath G A and Turchi C S 2011 Life cycle assessment of a parabolic trough concentrating solar power plant and the impacts of key design alternatives *Environ. Sci. Technol.* 45 2457–64

5.   Carroll J 2011 Worst drought in more than a century strikes Texas oil boom *Bloomberg* 13 June 2011 (available atwww.bloomberg. com/news/2011-06-13/worst-drought-in-more-than-a-century-threatens-texas-oil-natural-gas-boom.html)

6.   Clark C, Harto C, Sullivan J and Wang M 2011 *Water Use in the Development and Operation of Geothermal Power Plants* ANL/EVS/R-10/5 (Oak Ridge, TN: Argonne National Laboratory (ANL))

7.   DOE 1983 *Energy Technology Characterizations Handbook: Environmental Pollution and Control Factors* (Washington, DC: US Department of Energy (DOE))

8.   DOE 2006 *Energy Demands on Water Resources: Report to Congress on the Interdependency of Energy and Water*(Washington, DC: US Department of Energy (DOE))

9.   DOE 2012 *Sunshot Vision Study* (Washington, DC: US Department of Energy (DOE))

10. Dolan S L and Heath G A 2012 Life cycle greenhouse gas emissions of utility-scale wind power: systematic review and harmonization *J. Ind. Ecol.* 16 S136–54

11. EIA 2011a *Form EIA-860: Annual Electric Generator Report* (Washington, DC: US Department of Energy, Energy Information Administration (EIA))

12.   EIA 2011b *Form EIA-923 Power Plant Operations Report Instructions* (Washington, DC: US Department of Energy, Energy Information Administration (EIA))

13.   EIA 2011c *Review of Emerging Resources: US Shale Gas and Shale Oil Plays* (Washington, DC: US Department of Energy, Energy Information Administration (EIA))

14.   EPA 2011 *Plan to Study the Potential Impacts of Hydraulic Fracturing on Drinking Water Resources* EPA/600/R-11/122 (Washington, DC: US Environmental Protection Agency (EPA))

15.   Fthenakis V and Kim H C 2010 Life-cycle uses of water in US electricity generation *Renew. Sustain. Energy Rev.* 142039–48

16.   GAO 2009 *Energy-Water Nexus: Improvements to Federal Water Use Data Would Increase Understanding of Trends in Power Plant Water Use* GAO-10-23 (Washington, DC: US Government Accountability Office (GAO))

17.   GAO 2012 *Oil and Gas: Information on Shale Resources, Development, and Environmental and Public Health Risks*GAO-12-732 (Washington, DC: US Government Accountability Office (GAO))

18.   Genesee County Economic Development Center 2011 *SEQRA Review of Western New York Science & Technology Advanced Manufacturing Park (Stamp), Industry Requirements and Environmental, Health & Safety Report, Generic Environmental Impact Statement* (Batavia, NY: Genesee County Economic Development Center)

19.   Gerbens-Leenes P W, Hoekstra a Y and van der Meer T 2009 The water footprint of energy from biomass: a quantitative assessment and consequences of an increasing share of bio-energy in energy supply *Ecol. Econ.* 68 1052–60

20.   Gleick P H 1994 Water and energy *Annu. Rev. Energy Environ.* 19 267–99

21.   Grubert E A, Beach F C and Webber M E 2012 Can switching fuels save water? a life cycle quantification of freshwater consumption for texas coal- and natural gas-fired electricity *Environ. Res. Lett.* 7 045801

22.   Heath G A and Mann M K 2012 Background and reflections on the life cycle assessment harmonization project *J. Ind. Ecol.* 16 S8–S11

23. Hsu D D, O'Donoughue P, Fthenakis V, Heath G A, Kim H C, Sawyer P, Choi J-K and Turney D E 2012 Life cycle greenhouse gas emissions of crystalline silicon photovoltaic electricity generation: systematic review and harmonization *J. Ind. Ecol.* 16 S122–35

24. Huertas A 2007 *Rising Temperatures Undermine Nuclear Power's Promise. Union of Concerned Scientists Backgrounder* (Washington, DC: Union of Concerned Scientists)

25. Hutson S S, Barber N L, Kenny J F, Linsey K S, Lumia D S and Maupin M A 2004 *Estimated Use of Water in the United States in 2000 (US Geological Survey Circular* vol 1268) (Reston, VA: US Geological Survey)

26. IEA 2012 *Golden Rules for a Golden Age of Gas: World Energy Outlook Special Report on Unconventional Gas* WEO-2012 (Paris: International Energy Agency (IEA))

27. Inhaber H 2004 Water use in renewable and conventional electricity production *Energy Sources* 26 309–22

28. JISEA (Joint Institute for Strategic Energy Analysis) 2012 *Natural Gas and the Transformation of the US Energy Sector: Electricity* NREL/TP-6A50-55538 ed J Logan *et al* (Golden, CO: US Department of Energy, National Renewable Energy Laboratory (NREL))

29. Kenny J F, Barber N L, Hutson S S, Linsey K S, Lovelace J K and Maupin M A 2009 *Estimated Use of Water in the United States in 2005 (US Geological Survey Circular* vol 1344) (Reston, VA: US Geological Survey)

30. Kim H C, Fthenakis V, Choi J-K and Turney D E 2012 Life cycle greenhouse gas emissions of thin-film photovoltaic electricity generation: systematic review and harmonization *J. Ind. Ecol.* 16 S110–21

31. Kozubal E and Kustcher C 2003 Analysis of a water-cooled condenser in series with an air-cooled condenser for a proposed 1-MW geothermal power plant *Int. Collaboration for Geothermal Energy in the Americas: Proc. of the Geothermal Resources Council 2003 Annual Mtg (Morelia, Michoacan, Mexico, Oct. 2003)* (Davis, CA: Geothermal Resources Council) pp 587–591

32. Lustgarten A 2009 Frack fluid spill in Dimock contaminates stream, killing fish *ProPublica* 21 September 2009 (available

at   www.propublica.org/article/frack-fluid-spill-in-dimock-contaminates-stream-killing-fish-921)

33.   Macknick J, Newmark R, Heath G and Hallett K C 2011 *A Review of Operational Water Consumption and Withdrawal Factors for Electricity Generating Technologies* (Golden, CO: National Renewable Energy Laboratory) (www.nrel.gov/docs/fy11osti/50900.pdf)CrossRef

34.   Macknick J, Newmark R, Heath G and Hallett K 2012 Operational water consumption and withdrawal factors for electricity generating technologies: a review of existing literature *Environ. Res. Lett.* 7 045802

35.   McMahon J E and Price S K 2011 Water and energy interactions *Annu. Rev. Environ. Resources* 36 163–91

36.   Mielke E, Diaz Anadon L and Narayanamurti V 2010 Water consumption of energy resource extraction, processing, and conversion *Discussion Paper 2010-15* (Cambridge, MA: Energy Technology Innovation Policy Research Group, Belfer Center for Science and International Affairs, Harvard Kennedy School, Harvard University)

37.   MIT 2007 *The Future of Coal: Options for a Carbon-Constrained World* (Cambridge, MA: Massachusetts Institute of Technology (MIT))

38.   Moomaw W, Burgherr P, Heath G, Lenzen M, Nyboer J and Verbruggen A 2011 Annex II: methodology *IPCC Special Report on Renewable Energy Sources and Climate Change Mitigation* ed O Edenhofer, R Pichs-Madruga, Y Sokona, K Seyboth, P Matschoss, S Kadner, T Zwickel, P Eickemeier, G Hansen, S Schlömer and C von Stechow (New York: Cambridge Univeristy Press) pp 974–1000

39.   Mudd G M and Diesendorf M 2009 Response to Comment on 'Sustainability of uranium mining and milling: toward quantifying resources and eco-efficiency' *Environ. Sci. Technol.* 43 3969–70

40.   NETL 2009a *Existing Plants, Emissions and Capture—Setting Water-Energy R&D Program Goals DOE/NETL-2009/1372* (Pittsburgh, PA: US Department of Energy, National Energy Technology Laboratory (NETL))

41.   NETL 2009b *Impact of Drought on US Steam Electric Power Plant Cooling Water Intakes and Related Water Resource Management*

Issues DOE/NETL-2009/1364 (Pittsburgh, PA: US Department of Energy, National Energy Technology Laboratory (NETL))

42. NETL 2010a *Life Cycle Analysis: Existing Pulverized Coal (EXPC) Power Plant* DOE/NETL-403-110809 (Pittsburgh, PA: US Department of Energy, National Energy Technology Laboratory (NETL))

43. NETL 2010b *Life Cycle Analysis: Natural Gas Combined Cycle (NGCC) Power Plant* DOE/NETL-403-110509 (Pittsburgh, PA: US Department of Energy, National Energy Technology Laboratory (NETL))

44. NETL 2010c *Life Cycle Analysis: Supercritical Pulverized Coal (SCPC) Power Plant* DOE/NETL-403-110609 (Pittsburgh, PA: US Department of Energy, National Energy Technology Laboratory (NETL))

45. NETL 2010d *NETL Life Cycle Inventory Data-Unit Process: Natural Gas Sweetening* (Pittsburgh, PA: US Department of Energy, National Energy Technology Laboratory (NETL))

46. NETL 2012a *Role of Alternative Energy Sources: Nuclear Technology Assessment* DOE/NETL-2011/1502 (Pittsburgh, PA: US Department of Energy, National Energy Technology Laboratory (NETL))

47. NETL 2012b *Role of Alternative Energy Sources: Solar Thermal Technology Assessment* DOE/NETL-2012/1532 (Pittsburgh, PA: US Department of Energy, National Energy Technology Laboratory (NETL))

48. Noble Energy Inc. and CSU 2012 *Lifecycle Analysis of Water Use and Intensity of Noble Energy Oil and Gas Recovery in Wattenberg Field of Northern Colorado* (Fort Collins, CO: Noble Energy, Inc., and Colorado State University (CSU))

49. O'Donoughue P R, Dolan S L, Heath G A and Vorum M 2012 Life cycle greenhouse gas emissions from natural gas-fired electricity generation: systematic review and harmonization *J. Ind. Ecol.* submitted

50. Passwaters M 2011 *Texas Drought Starts to Pinch Barnett Producers. September 1, 2011* (Charlottesville, VA: SNL Financial LC) (available at www.snl.com/InteractiveX/ArticleAbstract. aspx?id=13250750)

51. Saathoff K 2011 *Grid Operations and Planning Report. Presentation to ERCOT Board of Directors Mtg* (available at www.ercot.com/content/meetings/board/keydocs/2011/1018/Item_04e_-_Grid_Operations_and_Planning_Report.pdf)

52. Sahm A, Gray A, Boehm R and Stone K 2005 Cleanliness maintenance for an amonix lens system *Proc. ISEC2005 2005 Int. Solar Energy Conf. (Orlando, FL, Aug.)* ISEC2005-76036

53. Schneider E, Carlsen B and Tavrides E 2010 *Measures of the Environmental Footprint of the Front End of the Nuclear Fuel Cycle* INL/EXT-10-20652 (Idaho Falls, ID: US Department of Energy (DOE), Idaho National Laboratory)

54. Solar Millennium LLC 2008 *Updated Plan of Development, Amargosa—Farm Road Solar Project* NVN-84359 (Las Vegas, NV: US Department of Interior, Bureau of Land Management (BLM))

55. Solley W B, Pierce R R and Perlman H A 1998 *Estimated Use of Water in the United States in 1995 (US Geological Survey Circular* vol 1200*)* (Reston, VA: US Geological Survey)

56. Stone K C, Hunt P G, Cantrell K B and Ro K S 2010 The potential impacts of biomass feedstock production on water resource availability *Bioresource Technol.* 101 2014–25

57. Tolba M K 1985 *The Environmental Impacts of Production and Use of Energy. Part IV—the Comparative Assessment of the Environmental Impacts of Energy Sources, Phase 1—Comparative Data on the Emissions, Residuals and Health Hazards of Energy Sources. Energy Report Series* ERS 14-85 (Nairobi: United Nations Environment Programme)

58. Torcellini P, Long N and Judkoff R 2003 *Consumptive Water Use for US Power Production* NREL/CP-550-35190 (Golden, CO: US Department of Energy, National Renewable Energy Laboratory (NREL))

59. Turchi C S, Wagner M J and Kustcher C F 2010 *Water Use in Parabolic Trough Power Plants: Summary Results from WorleyParsons' Analyses* NREL/TP-5500-49468 (Golden, CO: US Department of Energy, National Renewable Energy Laboratory (NREL))

60. TWDB 2003 *Power Generation Water Use in Texas for the Years 2000 Through 2060: Final Report* (Austin, TX: Texas Water Development Board (TWDB))

61. TWDB 2012 *Water for Texas 2012 State Water Plan* (Austin, TX: Texas Water Development Board (TWDB))

62. Ward K Jr 2010 Environmentalists urge tougher water standards *The Charleston Gazette* 19 July 2010 (available athttp://sundaygazettemail.com/News/201007190845)

63. Warner E S and Heath G A 2012 Life cycle greenhouse gas emissions of nuclear electricity generation: systematic review and harmonization *J. Ind. Ecol.* 16 S73–92

64. Whitaker M, Heath G A, O'Donoughue P and Vorum M 2012 Life cycle greenhouse gas emissions of coal-fired electricity generation: systematic review and harmonization *J. Ind. Ecol.* 16 S53–72

65. Wilson W, Leipzig T and Griffiths-Sattenspiel B 2012 *Burning Our Rivers: The Water Footprint of Electricity* (Portland, OR: River Network, Rivers, Energy and Climate Program)

66. Yang X and Dziegielewski B 2007 Water use by thermoelectric power plants in the United States *J. Am. Water Resources Assoc.* 43 160–9

Chapter 2

The Banana Code—
Natural Blend Processing
in the Olfactory Circuitry of
Drosophila melanogaster

Marco Schubert, Bill S. Hansson, and Silke Sachse

Department of Evolutionary Neuroethology, Max Planck Institute for Chemical Ecology, Jena, Germany

ABSTRACT

Odor information is predominantly perceived as complex odor blends. For *Drosophila melanogaster* one of the most attractive blends is emitted by an over-ripe banana. To analyze how the fly's olfactory system processes natural blends we combined the experimental advantages of gas chromatography and functional imaging (GC-I). In this way, natural banana compounds were presented successively to the fly antenna in close to natural occurring concentrations. This technique allowed us to identify the active odor components, use these compounds as stimuli and measure odor-induced Ca^{2+} signals in input and output neurons of the *Drosophila* antennal lobe (AL), the first olfactory neuropil. We demonstrate that mixture interactions of a natural blend are very rare and occur only at the AL output level resulting in a surprisingly linear blend representation. However, the information regarding single components is strongly modulated by the olfactory circuitry within the AL leading to a higher similarity between the representation of individual components and the banana blend. This observed modulation might tune the olfactory system in a way to distinctively categorize odor components and improve the detection of suitable food sources. Functional GC-I thus enables analysis of virtually any unknown natural odorant blend and its components in their relative occurring concentrations and allows characterization of neuronal responses of complete neural assemblies. This technique can be seen as a valuable complementary method to classical GC/electrophysiology techniques, and will be a highly useful tool in future investigations of insect-insect and insect-plant chemical interactions.

INTRODUCTION

The natural environment displays a myriad of vital cues coded in complex odor blends, which often are composed of a large number of single odor components. Information processing of simultaneous input regarding several different odor compounds forming a specific and behaviorally relevant representation is so far poorly understood. Hereby, a question of general importance arises: Does the olfactory system process and encode simultaneously occurring components as blend-specific information? And does this representation evolve over the different levels of olfactory processing? We addressed these questions by analyzing physiological responses to a natural odor blend and its single odor components in the antennal lobe (AL) of the vinegar fly *Drosophila melanogaster*.

Drosophila detects odor molecules with two olfactory organs, the maxillary palps and the antennae. Different types of olfactory sensilla house olfactory sensory neurons (OSNs) carrying different types of odorant receptors (ORs) (Hallem and Carlson, 2006; Vosshall and Stocker, 2007; Hansson et al., 2010). OSNs can either be narrowly tuned or respond to a broad range of structurally similar odor ligands (De Bruyne et al., 2001; Hallem and Carlson, 2006; Pelz et al., 2006;

Stensmyr et al., 2012). From the antenna the information is conveyed to the ALs, the first relay station of the olfactory pathway (Figure 1A). Each group of OSNs, carrying the same type of OR, converge onto one or a few specific olfactory glomeruli (Gao et al., 2000; Vosshall et al., 2000; Couto et al., 2005; Fishilevich and Vosshall, 2005; Silbering et al., 2011). Each AL comprises about 50 glomeruli, which represent structural and functional units that shape and modulate the odor information on its way to higher processing centers (Laissue et al., 1999; Couto et al., 2005; Galizia and Sachse, 2010). Within the glomeruli, OSNs exchange information with local interneurons (LNs) and projection neurons (PNs) by excitatory and inhibitory synaptic crosstalk (Wilson, 2011). Since each OSN type targets its own specific glomerulus, the detection of odor molecules leads to a specific mosaic of glomerular activity patterns (Fiala et al., 2002; Ng et al., 2002; Wang et al., 2003; Silbering et al., 2008).

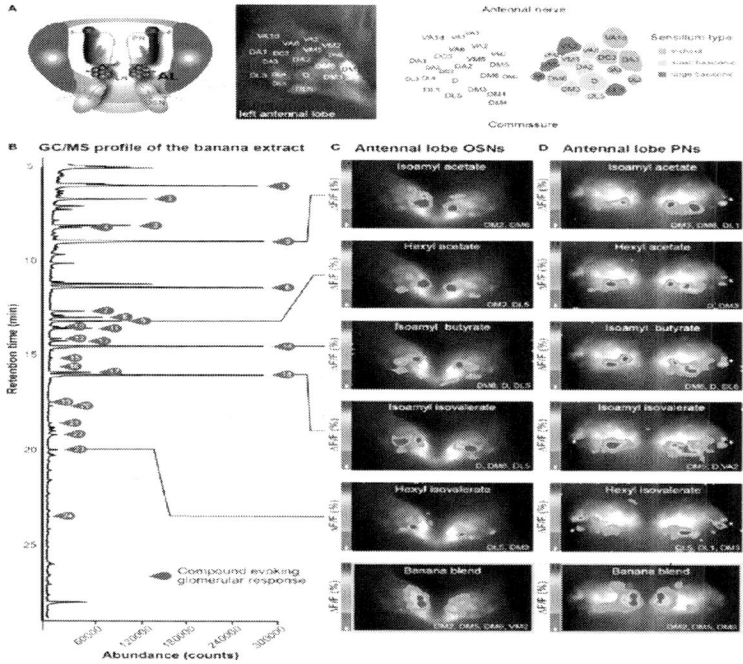

Figure 1: Neuronal activity patterns of banana compounds in the fly antennal lobe using combined gas chromatography and functional

imaging (GC-I). (A) *Left*, schematic illustrating the *Drosophila* olfactory system. Odor information detected by the antenna is conveyed by olfactory sensory neurons (OSN: red) to the antennal lobe (AL). OSNs converge in specific glomeruli and synapse onto local interneurons (LNs: gray) and projection neurons (PNs: green). *Middle*, baseline fluorescence of G-CaMP in axon termini of OSNs in the AL with anatomical identification of individual glomeruli. The antennal nerve enters the imaged region at the top; the AL commissure is at the bottom. *Right*, schematic AL map viewed from the angle used for imaging experiments. Colored glomeruli ($n = 17$) could reliably be identified; colors correspond to their sensillum input. (B) GC/MS profile of a banana extract revealing single compounds that compose the banana headspace extract (see Table 1 for peak identity of active compounds corresponding to the red number tags).(C,D) Pseudocolor rendering of Ca^{2+} responses of OSNs (C) and PNs (D) to different banana extract compounds and the complete banana blend. Images represent $F/F_0(\%)$ superimposed onto raw fluorescence images according to the scale bars on the left. Black lines serve to link the corresponding peaks (B) to the evoked activity patterns. White asterisk shows cell body activity in the lateral cell clusters. The most active glomeruli are indicated in the lower right corner.

When two odors are processed simultaneously by the olfactory system, odor mixture interactions might occur. These can result in either suppression or synergism leading to a reduced or an enhanced mixture response compared to single component responses (Akers and Getz, 1993;Duchamp-Viret et al., 2003; Deisig et al., 2006; Silbering and Galizia, 2007; Rospars et al., 2008;Kuebler et al., 2011; Münch et al., 2013). Most recent physiological studies on odor mixture processing have mainly focused on binary or quaternary mixtures with monomolecular synthesized odor compounds because of application advantages (Deisig et al., 2006; Silbering and Galizia, 2007;Grossman et al., 2008; Fernandez et al., 2009; Deisig et al., 2010). In order to analyze neuronal processing of a complex naturally occurring mixture and subsequently to identify its individual odor components, we combined the experimental advantages of gas chromatography with functional imaging, subsequently called functional GC-I, as previously established for the mouse olfactory system (Lin et al., 2006). Here, we examine one of the most attractive food sources and breeding places for *Drosophila melanogaster*, an over-ripe banana (Sturtevant,

1921; Lachaise and Silvain, 2004), in its natural composition. Using functional GC-I enabled us to capture and identify potentially relevant odor components of a banana headspace extract. In addition this technique allowed us to use these components as stimuli in close to natural relative proportions, and monitor odor-induced Ca^{2+} dynamics *in vivo*. To do so, we expressed the genetically encoded Ca^{2+} indicator G-CaMP at different processing levels of the *Drosophila* AL. We show that the banana extract consists of 24 active compounds that induce clear neuronal activity in AL input and output neurons. We demonstrate that mixture interactions within a natural banana blend are very rare and occur only at the AL output level resulting in a surprisingly linear blend representation. We further show, however, that individual glomerular responses are significantly modified by the neural circuitry in the AL from the input to the output level resulting in a modulated odor representation. This modulation leads to a higher similarity between the representations of individual components in relation to the complete banana blend. Such processing mechanism might tune the olfactory system in a way to categorize key components with their naturally occurring odor source to enhance the detection of suitable food sources.

MATERIALS AND METHODS

Animals

We used 6–10 days old female vinegar flies (*Drosophila melanogaster*) raised on conventional cornmeal-agar-molasses medium under L:D/12:12, *RH*=70% and 25°C. Transgenic lines used: Orco-GAL4 (Larsson et al., 2004), GH146-GAL4 (Stocker et al., 1997), UAS-G-CaMP1.6 (Nakai et al., 2001; Wang et al., 2003). Flies were dissected for optical imaging as described (Stökl et al., 2010;Strutz et al., 2012). Briefly, flies were anesthetized on ice, fixed with the neck onto a Plexiglas stage using a copper plate (Athene Grids). The head was glued at the stage with colophony resin (Royal Oak, Rosinio) and the antennae were gently pulled forward with a fine metal wire (H.P. REID co. inc., USA). Polyethylene foil was attached on the head and sealed to the cuticle with two-component silicone (KwikSil, WPI). A small hole was cut through the foil and cuticle. Immediately after opening of the head, the brain was bathed with Ringer solution (130 mM NaCl, 5 mM KCl, 2 mM $MgCl_2$, 2 mM $CaCl_2$, 36 mM sucrose, 5 mM Hepes, [pH 7.3]). Removal of trachea and glands allowed optical access to the ALs.

Optical Imaging

We used a Till Photonics imaging system with an upright Olympus microscope (BX51WI) equipped with a 20 × Olympus objective (XLUM Plan FL 20x/0.95W). A Polychrome V provided light excitation (475 nm) and a filter set ensured passage of only relevant wavelengths (excitation: SP500, dicroic: DCLP490, emission: LP515). The emitted light was captured by a CCD camera (Sensicam QE, PCO AG) with a symmetrical binning of 4 (1.25 × 1.25 µm/pixel). For each measurement a series of 300 frames was taken (1 Hz, GC-I run time 3.5–8.5 min). A low sample rate of 1 Hz prevented the fluorescent Ca^{2+} sensor from bleaching over the 300 images taken. To assure that we did not lose any information by the low imaging sample rate of 1 Hz we also tested a 2 Hz frequency showing that no additional response peaks were registered (data not shown).

Data Analysis

All imaging data were analyzed using custom software written in IDL (ITT Visual Information Solutions). For anatomical identification of glomeruli we compared the glomerular organization of the ALs with an available standard atlas (Laissue et al., 1999) as we have previously described in detail (Stökl et al., 2010). For data analyses, the activity of individual glomeruli was taken as an area of 5 × 5 pixels per glomerulus. A bleaching correction was applied for each frame (300 frames per imaging sequence) by subtracting the median fluorescence from each pixel. An automated movement correction compensated for movement artifacts between frames during the imaging sequence. To achieve a comparable standard for the calculation of the relative fluorescence changes (F/F_0), we defined the background fluorescence (F_0) as the mean of 10 successive frames before the stimulation with the extract components. This background was then subtracted for each glomerulus during the whole sequence of 300 frames, so that basal fluorescence has been normalized to zero. The calcium responses of each identified glomerulus were synchronized to the GC banana profile data by using isoamyl acetate and butyl butyrate peaks as prominent orientation points. A data matrix was generated for the fluorescence changes of each identified glomerulus over each of the 300 frames imaged. Glomerular responses were normalized within each animal to the strongest glomerulus response measured which was set to 100%. After identification of all 70 odor components of the banana extract, we included in our analysis only those fluorescent changes that corresponded to the component peaks in the GC run for each glomerulus. We then determined those components that induced a calcium signal resulting in 24 active components (Table 1) and used these for further analysis. Hence our data matrix represents population vectors that are defined by the identity of a glomerulus in one dimension and the calcium signal for each of the 24 odor components in the other dimension.

Table 1: Identified and physiological active banana compounds

Order of compounds	Retention time (min)	Kovats indices	Absolute peak height (MS/GC counts)	CAS	Compound
1	6.172	770	1985038	110-19-0	Isobutyl acetate
2	6.878	802	425162	105-..4	Ethyl butyrate
3	8.273	850	486014	62,3.	2-Pentyl acetate
4	8.398	8.	93711	108..-5	Ethyl isovalerate
5	9.127	878	7707050	123-92-2	Isoamyl acetate
6	11.537	956	3759896	539-.2	Isobutyl butyrate
7	12.799	996	2450746	109-21-7	Butyl butyrate
8	13.127	1007	.4566	589-59-3	Isobutyl isovalerate
9	13.352	1014	4112522	142-92-7	Hexyl acetate
10	13.619	1023	2718249	72237-3.	4-Hexenyl acetate
11	13.721	1026	3100127	60415-61.4	2-Pentyl butyrate
12	14.255	1044	2318335	5921-82-4	2-Heptyl acetate
13	14.381	1048	.0472	109-1.3	1-Butyl isovalerate
14	14.812	1062	13284855	10,27-4	Isoamyl butyrate
15	15.281	1077	373711	89155-38-4	2-Pentylvalerianate

16	18.827	1095	63619	n/a	MIX: n-pentyl butyrate/ n-butyl valerate
17	16.024	1101	2283078	2762...	Isoamyl 2-Methyl butyrate
18	16.261	1110	11475565	659-70-1	Isoamyl isovalerate
19	17.507	1152	181278	105-7.	Isobutyl hexanoate
20	17.565	1154	339344	2050-0.1	Isoamyl valerate
	18.686	1193	1282073	2639-63-6	Hexyl butyrate
22	19.292	1214	.2710	39026-94-3	2-Heptyl butyrate
23	20.081	1243	2062.3	10032-13-0	Hexyl isovalerate
24	23.677	1378	32112	n/a	Isomer of octenyl butyrate

Listed are the banana compounds following the retention times needed to be detected by the FID. Compounds were identified by Kovats retention indices Inon.isotherm, for temperature programmed methods.

Normalized responses within identified glomeruli were compared using Student's *t*-tests (unpaired two-tailed distribution). In order to analyze the proximity of our odor representations to the 24 active components and the banana blend in a putative neural space, we regarded each odor representation as a vector in a multidimensional space, in which each dimension is represented by a glomerulus. We used the relative fluorescence changes (F/F_0) in single frames (i.e., corresponding to each one of the 24 components) for each identified glomerulus and calculated the Euclidean distances between each single odor component and the banana blend to quantify the pattern proximity. Furthermore, we applied principal coordinates analysis including all glomeruli that we could identify at both processing levels ($n = 10$) in order to visualize the pattern similarity in a lower-dimensionality space formed of a subset of highest-variance components (Deisig et al., 2010). Statistical analyses were performed with the software GraphPad Instat and PAST.

Odor Extract

Banana extracts were produced from commercially available ripe bananas. Cut bananas including the skin were placed into an oven-bag (Toppits© Roasting-bags, www.toppits.de) which was perforated with air holes on one side and connected to a Super-Q filter (50 mg, Analytical Research Systems, Inc.) on the other side. A pump (Casella Apex lite) sucked banana odor laden air for 4 h with a constant flow rate (1 l/min.) over the filter, which was eluted with hexane (300 µl) afterwards and the extract was stored at −20°C until use. Silicon tubing and Teflon© connectors were used to avoid contamination. In control experiments extracts coming from bananas of different age (degree of fermentation) showed similar GC profiles in terms of individual components and only partially differences with respect to component concentration (data not shown). Extracts from older bananas typically provided higher concentrations of molecules emerging in the first half of the GC profile. Despite major concentration differences, the glomerular activity was qualitatively almost concentration independent for the used extracts.

Gas Chromatography

We injected 2 µl of banana extract into an Agilent 6890N GC (Agilent Technologies). Separated extract components were leaving the GC via a heated and flexible transfer line (GC outlet). The transfer line head was mounted with a Pasteur pipette in which the components were injected. A constant purified and humidified airstream carried the stimuli through the pipette to the fly antenna. For GC parameter control and data acquisition an external computer running the commercial software GC ChemStation (Agilent Technologies) was used. GC banana blend data collected during imaging (5 min, sample rate of 1200 Hz) were synchronized with the imaging data for detailed comparisons. Subsequent GC/MS (5975B inert XL MSD, Agilent Technologies) analysis was used for identification of all active components.

The Agilent 6890N GC (Agilent Technologies) was running the injector in splitless mode (250°C) using helium as a makeup/carrier gas which did not induce any glomerular responses. At the end of a HP-5 low/non-polar column (flow rate: 2 ml/min; column length: 30 m, inner diameter: 0.32 mm, inside coating: 0.25 µm, thick film of 5% phenyl methyl siloxane and 95% methyl siloxane) the sample stream was split in two parts (1:1), one leading to a flame ionization detector (FID, detector temp.: 310°C) and the other leaving the GC via a heated and flexible transfer line (GC outlet). During each run the GC oven and transfer line temperature was synchronized, ramped from 40°C (1 min) at 20°C/min to 300°C. The transfer line head at the end (300°C constantly) was mounted with a Pasteur pipette (length: 12cm) in which the separated stimuli components were injected via the GC outlet.

Odor Puff Stimulation

After each functional GC-I run the animals were exposed to odor puff stimulations with the banana extract, the solvent hexane and an air control, while glomerular AL responses were optically recorded. A stimulus controller (CS-55, Syntech) provided a continuous air flow (0.5 l/min) in which odor injection could be applied via two disposable

Pasteur pipettes. For odor stimulation the air stream switched from a blind Pasteur pipette to the stimulus pipette in which the filter paper was odor laden for 2 s. The banana extract was applied in the same concentration as the GC fractionated banana components to allow for subsequent comparison.

RESULTS

Neuronal Representation of Banana Odor Components

By combining the experimental advantages of gas chromatography and Ca^{2+} imaging, we measured the representation of single banana compounds and the complete banana blend at different levels of olfactory processing in the *Drosophila* olfactory system. The volatile collection of an over ripe banana was injected into a GC, where it was separated into more than 70 individual components (Figure 1B). In this way, banana odor components could be presented successively to the fly antenna in naturally occurring concentrations. An imaging capture rate of 1 Hz allowed us to measure the responses to each single component that emerged from the GC, since components were separated by at least 1 s. Identification of components that induced a significant increase in the intracellular calcium concentration ($[Ca^{2+}]_i$) in the AL was subsequently performed using GC-mass spectrometry (Table 1) and was well in line with compounds earlier identified in banana headspace extracts (Shiota, 1993; Jordán, 2001; Stensmyr et al., 2003).

Primarily, we measured the representation of single banana odor compounds and the complete banana blend in input neurons, i.e. the axonal terminals of OSNs in the AL (Figure 1C). Using the binary GAL4-UAS transcriptional system (Brand and Perrimon, 1993), we genetically expressed the Ca^{2+}-sensitive reporter G-CaMP (Nakai et al., 2001) in the majority of OSNs employing Orco-GAL4 (Wang et al., 2003). Since the AL morphology with its glomerular structure is invariant and clearly visible, we could identify individual glomeruli in every animal using the available 3D atlas of the *Drosophila* AL (Laissue et al., 1999). This identification enabled us to assign odor-evoked Ca^{2+}

responses to 17 glomeruli (52% of all glomeruli labeled by Orco-GAL4) and hence to correlate those to specific sensilla and OR types on the antenna (Figures1A,C) (Hansson et al., 2010). We observed significant odor-evoked Ca^{2+} responses to 24 out of the 70 banana extract compounds. Activation of OSNs by single banana components and by the complete blend resulted in specific combinatorial patterns of activated glomeruli.

Secondly, we examined the representation of the single banana compounds and the complete blend at the next processing level, the dendrites of AL output neurons. To achieve this, we expressed G-CaMP in the majority of PNs using the enhancer trap line GH146-GAL4 (Figure 1D) (Stocker et al., 1997). Similar to the OSN recordings, the odor-evoked responses could be reliably assigned to 15 identified glomeruli (41% of all glomeruli labeled by GH146-GAL4). Since GH146-GAL4 does not label glomeruli VM5 and VA6, these could not be characterized at the PN level. We observed specific odor-evoked Ca^{2+} responses for the same 24 banana compounds as detected during the OSN recordings. The complete banana blend induced a broad but, nevertheless, specific pattern of activated glomeruli at both processing levels (Figures 1C,D lowest panel).

Comparison between Input and Output Representation

In order to allow a comparison between the two processing levels, we synchronized the response profiles of OSNs and PNs and aligned them to the chromatograms (Figure 2) using characteristic component landmarks such as isoamyl acetate and isoamyl butyrate (#5 and #14 in Figure 1B). Notably, glomerulus-specific time traces of OSNs and PNs showed similar, but not identical odor response properties indicating that the odor-evoked responses are modulated from the input to the output level. Several glomeruli revealed a broad response profile while others, in particular glomeruli receiving input from trichoid sensilla, showed only sparse or no activity at all. To simplify the recorded Ca^{2+} dynamics, we quantified the odor-evoked response intensities for all identified glomeruli and summarized these as a heat map for each individual banana compound and the blend response (Figure 3). We excluded here glomeruli receiving input from trichoid sensilla, since they did not show any significant responses to the tested banana components. The most prominent AL responses were recorded in glomeruli DM2 and DM6. Interestingly, odors with a similar retention time and therefore similar chemical properties activated a similar combination of glomeruli confirming a previous imaging study in honeybees (Sachse et al., 1999). The banana blend itself evoked a very broad response pattern (Figure 3, last row). Comparison between OSN and PN response intensities shows again that the odor representations are different between the two processing levels.

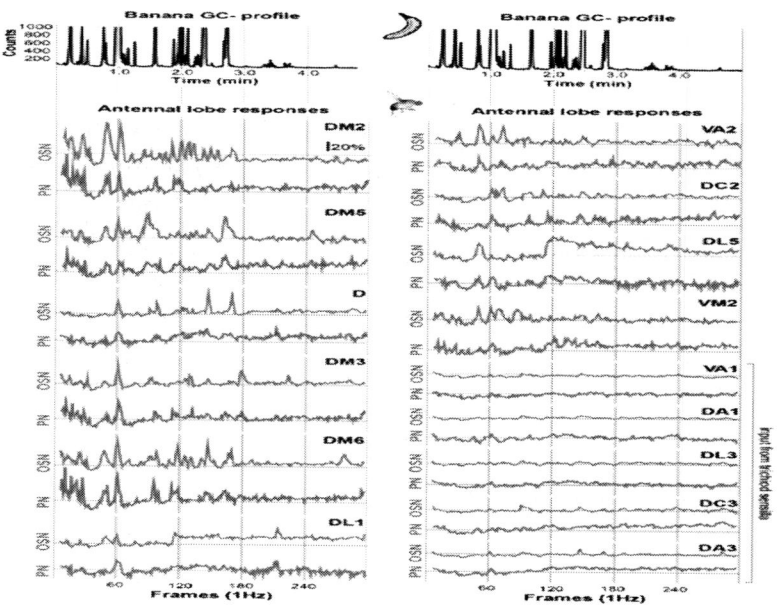

Figure 2: Glomerular response traces of input and output neurons to banana extract stimulation. *Top*, GC profile (black line) of the banana extract.*Below*, synchronized time courses of Ca^{2+} dynamics are shown for the OSN (orange traces) and PN level (green traces) averaged across five flies, respectively. Traces are given for all individual glomeruli that could be identified at both processing levels. The sample rate (frames/second) was 1 Hz. The whole measurement lasted for 5 min. Time traces represent the percentage of intensity changes compared to background activity (F/F).

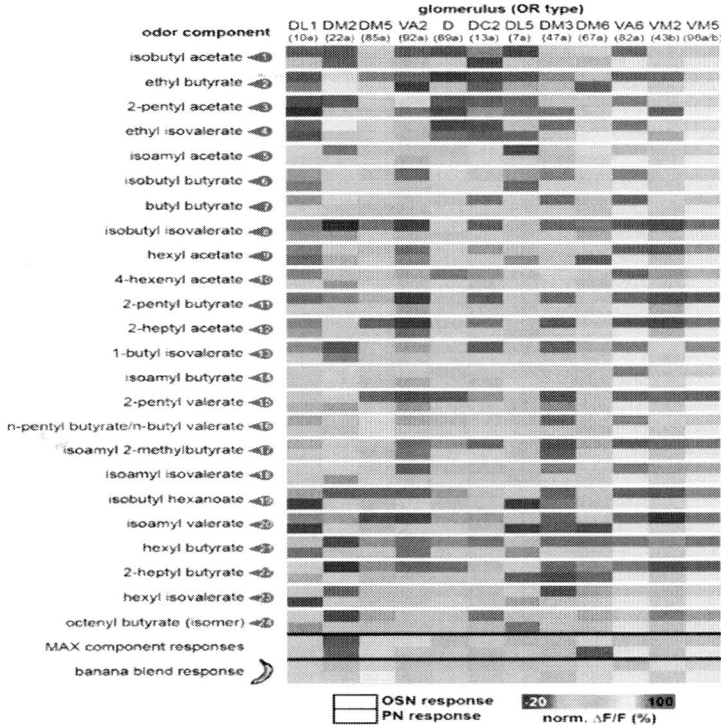

Figure 3: Functional map of odor-evoked glomerular activation to banana compounds. The odor responses of 12 glomeruli are shown for each identified banana compound at the OSN (upper box) and the PN level (lower box) as a heat map. Each data point is the median glomerular response from five flies. The last two rows represent glomerular responses to the measured complete banana blend and the strongest responses calculated for each component as a prediction for the banana blend response. For each individual glomerulus the corresponding odorant receptor input is given in brackets.

To further examine this difference, we compared in detail the response of the strongest activated glomeruli between OSNs and PNs. Figure 4 depicts comparisons of odor responses between the two processing levels for three exemplary banana extract components and the complete banana blend. As already visible in the heat map (Figure 3), several glomeruli showed significantly higher responses at the PN level than at the OSN level as shown, e.g., for the odor isobutyl acetate (Figure 4A). However, we also observed that some odors induced a significantly reduced PN response in comparison to the OSN response as shown for glomerulus DM2 (Figures 4B, C). Interestingly, when comparing the input and output glomerular responses to the complete banana blend, we found a significant reduction in the PN response of glomerulus VM2 (Figure 4D). The observed signal modulation between input and output neurons shows a strong diversity of odor information transfer in a glomerulus-specific manner. However, we only observed a significant signal modulation in 8% of glomerular responses, while the majority of glomeruli showed almost identical signals between the two processing levels.

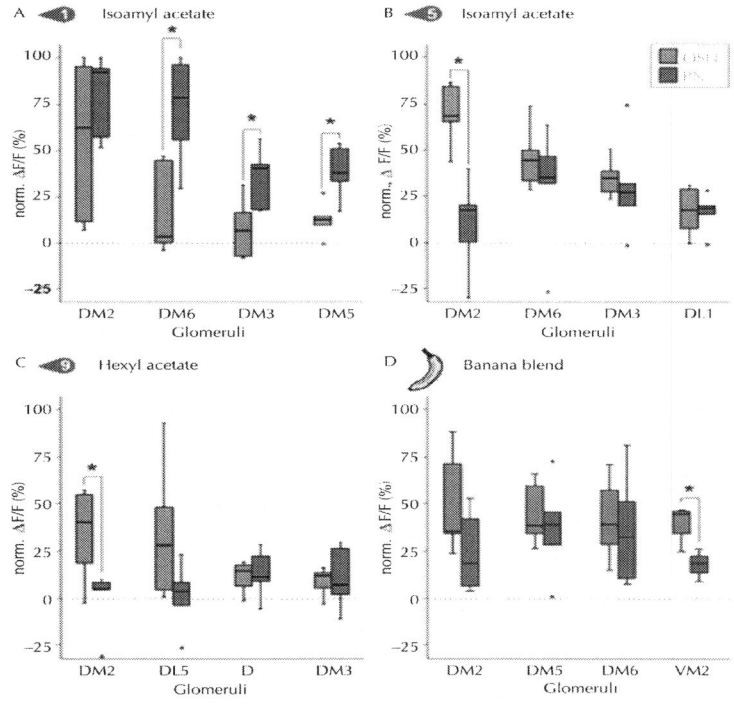

Figure 4: Odor response modulation between OSN and PN level. Examples of glomerular responses to three identified banana compounds (A–C) and the banana blend (D) are shown as a comparison between the OSN (orange) and PN (green) level for the four most active glomeruli per component. Box plots here and in Figure 5 represent the median value (horizontal line inside the box), the interquartile range (height of

the box, 50% of the data are within this range) and the minimum and maximum value (whiskers) of each experimental group. Circles depict outliers with values that were more than 1.5 times the interquartile range from the lower or upper quartile. Fluorescence values represent the average percentage of intensity changes compared to background activity (F/F_0, $n = 5$). Responses were normalized to highest calcium response in each animal over all odors before averaging. Significant differences are indicated with asterisks (*$p < 0.05$; unpaired t-test).

Linearity of Blend Representation

We next analyzed if the banana odor blend was linearly represented in the AL as predicted from the glomerular activation patterns induced by single components, or if the AL network was modulating the blend response to something different than predicted, implying non-linear blend effects. Since the concentration of the single components in the GC run is approximately equal as during the puff stimulation, we expect that the individual glomerular responses to the blend should be as strong as the maximal glomerular response to the single odor components (MAX component response). Interestingly, when we used this rather conservative approach to calculate the blend response, we could very well predict the actual measured blend response (Figure 3, last two rows). This observation is further supported by a direct comparison between the maximal component responses and the blend response which reveals no significant differences for any of the glomeruli measured at their input site (Figure 5A). The same analysis at the PN level shows a similar picture: The responses of most glomeruli did not differ between blend and single component stimulation except for glomerulus DM2 whose activity was significantly reduced during the actual blend application (Figure5B). Hence, blend interactions of the banana blend are rare and occur only at the AL output level leading to a surprisingly linear blend representation.

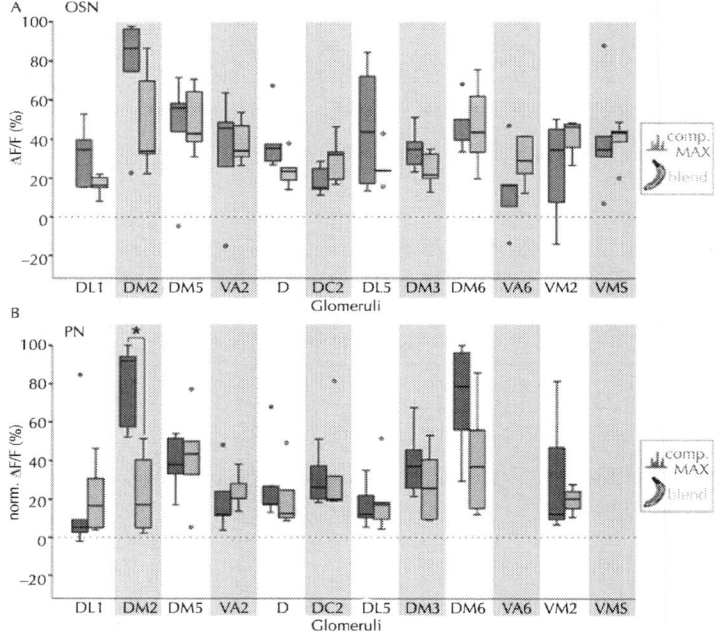

Figure 5: Comparison between individual compound and banana blend representation. (A) Orange boxplots represent the strongest OSN responses to any of the extract compounds, while yellow boxplots represent the response to the banana blend in the corresponding glomeruli. Differences between boxplots were not significantly different (unpaired *t*-test, *n* = 5). (B) Same analysis as in (A) at the PN level. Dark green boxplots represent the strongest PN component responses, whereas light green boxplots represent blend responses. Glomeruli VA6 and VM5v are not labeled by GH146-GAL4 and could therefore not be analyzed at the PN level. Glomerulus DM2 shows a significant lower response to the blend than to the strongest single component (*$p < 0.05$; unpaired *t*-test, *n* = 5).

Banana Blend Representation is modulated between Processing Levels

In order to analyze if a single odor component could be as representative as an over-ripe banana to a vinegar fly, we determined which one of the single components represented best the banana blend. To judge the similarity we calculated the Euclidean distances between components and blend response patterns and identified components producing the most similar response patterns compared to the banana evoked pattern (Table 2). Both OSN and PN similarity rankings include similar key components as 2-pentyl acetate (#3), ethyl isovalerate (#4), isoamyl acetate (#5) and isoamyl butyrate (#14). Interestingly, when we compared the Euclidean distances between the blend representation and the single components we observed that these were significantly lower at the PN level than at the OSN level (on average 0.89 for OSNs versus 0.63 for PNs; ***$p < 0.001$, paired T-test, Table 2, Figure 6A). This modulation leads to a higher similarity between the representations of individual components in relation to the complete banana blend.

Table 2: Component and blend similarity

Order of compounds	OSN Ed	PN Ed	Compound
5	52.41	47, 48	IsoanM acetate
14	65, 28	45, 67	ISOaMyl butyrate
3	70, 87	72, 35	2-Pentyl acetate
4	71, 90	88	Ethyl isovalerate
7	73, 23	45, 46	Butyl butyrate
10	79,85	66,44	4-Hexenyl acetate
18	81, 91	70, 98	Isoamyl isovalerate
16	82, 65	42,16	Mix: n-pentyl butyrate/n-butyl valerate
	83,12	51, 85	Isobutyl acetate
	85.41	73,04	Hexyl acetate
2	89,16	61,96	Ethyl butyrate
6	89, 30	58, 31	Isobutyl butyrate
17	90, 30	63, 05	Isoan* 2-methyl butyrate
11	96, 98	69,11	2-Pentyl butyrate
15	97,18	74, 22	2-Pentylvalerianate
	98, 29	67. 69	2-Heptyl acetate
13	98, 55	78, 99	1-Butyl isovalerate

23	98, 78	61,80	Hex,4 isovalerate
24	103, 67	53, 07	Isomer of octenyl butyrate
20	105, 42	66, 96	IsoariM valerate
19	105, 90	71, 49	Isobutyl hexanoate
21	107, 48	74, 75	Hexyl butyrate
22	108, 55	73	2-Heptyl butyrate
	108,86	68,54	Isobutyl isovalerate

Lineup of the Euclidean distances fEdl between component and blend responses.

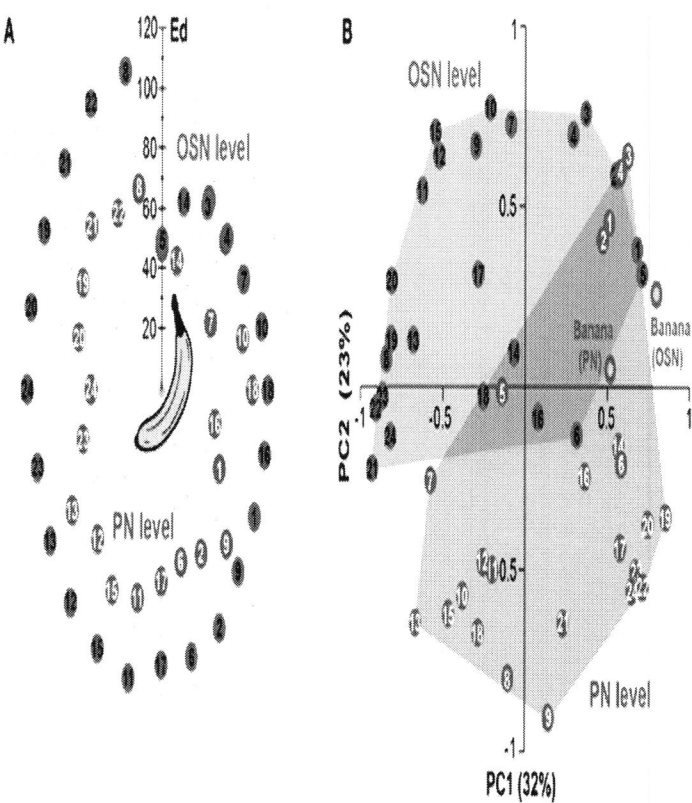

Figure 6: Odor component and blend similarity. (A) Similarity between single components and the banana blend responses at the OSN (orange number

tags) and PN (green number tags) level. The Euclidean distances (*Ed*) between component and blend responses were calculated and represented as the distance between the number tags and the banana center in a polar plot. Short distances between the tags and the center represent high, longer distances reveal low similarities. The similarities between component and blend representations are lower for all, except three components, at the PN level than at the OSN level. (B) Principal coordinates analysis of the individual component and blend responses at the OSN (orange) and PN (green) level. The odor representations of the two processing levels form significantly distinct clusters (***$p < 0.001$, One-Way ANOSIM, Bray-Curtis). The representation of the banana blend is located among the single component representations at the PN level, while it is outside of the cluster at the OSN level.

Next, we applied principal coordinates analysis to visualize the odor evoked ensemble activity to all banana compounds in relation to the blend pattern at both processing levels and found three main results (Figure 6B): First, the odor-evoked responses of the two levels form significantly distinct clusters supporting the observed activity pattern modulation between OSNs and PNs (***$p < 0.001$, One-Way ANOSIM, Bray-Curtis). Second, the different component representations of OSNs and PNs are spread over a similar sized odor space. And third, the blend representation appears outside of the cluster at the OSN level, while it is located among the single component representations at the PN level. Hence, the representation of individual banana components is shifted toward the blend representation at the output level. This change in physiological representation is leading to a higher similarity between the banana components in relation to the banana blend.

DISCUSSION

With the functional GC-I technique we established an experimental tool that solves the dilemma of using either a natural odor source (without knowing its exact component composition) or a synthetic odor blend (without knowing the natural concentrations) as an experimental stimulus. A fundamental question appearing when comparing functional GC-I runs and conventional blend stimulation is the use of comparable concentrations in both situations. Since we used identical extract concentrations that induced comparable glomerular response intensities for both odor applications, we conclude that both stimulations provided comparable stimulus concentrations to the animal's antenna. Another critical point is to verify that the animal was not adapted during the GC-I measurement. The single banana components were separated by at least 1 s when they emerged from the GC column, resulting in an inter-stimulus-interval (ISI) of 1 Hz. Since it has been shown that even PNs can reliably follow an ISI of up to 2 Hz without being adapted (Brown et al., 2005), and since we applied very low odor concentrations, we assume that the flies were not adapted during the GC-I recording.

Signal Modulation during the Transition from Input to Output Neurons

Our finding that some PN component responses were significantly reduced, while others were increased compared to the response intensity of OSNs (Figures 3, 4) proposes an interplay of inhibitory and excitatory processes caused by the neuronal network within the AL. The neuronal substrate for glomerulus-specific modulation is provided by inhibitory and excitatory LNs that have been characterized and suggested to be involved in the processing mechanisms of the AL (Wilson and Laurent, 2005; Shang et al., 2007; Okada et al., 2009; Chou et al., 2010; Seki et al., 2010). Lateral inhibition, accomplished by inhibitory LNs, has been shown to provide gain control which is defined as a negative feedback loop to keep the AL output in a given range (Olsen et al., 2010). The network of excitatory LNs is providing neuronal excitation between different glomeruli via cholinergic synapses and is assumed to improve odor detection at low intensities (Wilson, 2011).The signal modulation from the input to the output level that we observed in our study for a subset of glomeruli, most likely indicates gain control as well as an increased odor response specificity, allowing for improved odorant component and compound identification and discrimination. Interestingly a previous study by Bhandawat et al. showed that non-linear transformation of olfactory information led to signal broadening in PNs compared to an equivalent number of OSNs using electrophysiological techniques (Bhandawat et al., 2007). Weak OSN input was found to be amplified at the PN level, while strong input was not. Although, seemingly contradictive to our results, our work had the advantage of accounting simultaneously for activity in all OSNs and PNs accessible to our optophysiological technique. Consequently, we were able to investigate and compare more natural proportions of the neuronal populations of OSNs and PNs taking the strong convergence of the sensory input into account. Thus, our study adds complementary information since we investigated the neuronal ~10:1 (10 OSN synapse in average onto 1 PN) relationship between the OSN and PN level. Bhandawat et al.'s study showed stronger PN than OSN responses in 7 glomeruli in response to 18 odors comparing a similar number of OSNs and PNs (1:1). Analyzing the responses of 12 glomeruli (i.e. ~37% of all OSNs labeled by Orco-GAL4) in response to 24 odors of a natural odor source and in natural concentrations, we found that all kinds of inhibitory and excitatory network effects occurred when comparing the OSN with the PN level. It is thus impossible to provide a general rule regarding broadening or sharpening effects during signal transition between OSNs and PNs. Both processes do indeed seem to action in parallel. This observation is well in line with a previous study by Silbering et al. providing evidence for a complex and diverse processing mechanism across different glomeruli in the fly AL (Silbering et al., 2008).

Mixture Interaction Occurs only at the Output Level

In a purely linear model the blend representation would be predicted by a linear sum of the blend components. We used a conservative approach for identifying non-linear blend effects by comparing the response to the banana blend with the response to the strongest individual component for each specific glomerulus (Figure 5) (Deisig et al., 2006). The compound that induces the highest response should also represent the compound exhibiting the highest physiological salience in this glomerulus after blend stimulation. Blend responses lower than the response elicited by the most salient compound would therefore indicate mixture suppression (Silbering and Galizia, 2007). This conservative approach does not allow conclusions regarding synergistic effects which, however, have been shown to be exceedingly rare (Akers and Getz, 1993; Tabor et al., 2004; Silbering and Galizia, 2007). Interestingly, our comparison revealed no significant differences at the OSN level, while we found a significant effect of non-linear interactions in the DM2 glomerulus at the PN level. This result is well in line with the study by Silbering and Galizia (2007) showing that the representation of mixtures in *Drosophila* at the OSN level could rather be predicted from the response pattern of the single components, while mixture responses in PNs revealed strong mixture interactions. This is most likely due to the fact that PN responses are strongly modulated by interglomerular inhibition deriving from a glomerulus-specific network of inhibitory LNs (Wilson and Laurent, 2005; Silbering and Galizia, 2007). In addition, our observed linearity in OSN blend processing has earlier been reported in studies of numerous animal species (Tabor et al., 2004; Deisig et al., 2006; Lin et al., 2006; Carlsson et al., 2007). AL input neurons thus represent a linear blend information assembly, only being tuned by the dose response relationship of individual ORs. Well in line with our study is a recent study by Münch et al. that investigated mixture interactions to binary mixtures of banana compounds in the periphery of the fly olfactory system by performing calcium imaging of Or22a-expressing OSNs on the antenna (Münch et al., 2013) – an OSN population that targets glomerulus DM2 (Couto et al., 2005; Fishilevich and Vosshall, 2005). Münch et al. observed that mixture responses are hypoadditive, i.e. the mixture response was equal to the stronger component confirming our findings.

Interestingly, we could demonstrate that also at the level of the output neurons mixture interactions were surprisingly rare, which might be partially attributed to our conservative analysis. Although other studies have shown that global inhibitory network effects have increasing influence on blend interactions with the number of blend components (Deisig et al., 2006, 2010; Silbering and Galizia, 2007), this might only be true for synthetic mixtures and do not account for naturally occurring blends.

Similarity Shifts between Representations of the Blend and its Single Components

The similarity between glomerular activation patterns for all banana components was compared to patterns elicited by the complete banana blend. Assuming that distances between glomerular odor representations correlate with behavioral discrimination (Guerrieri et al., 2005), components with the highest similarity should be perceived as connatural as the banana blend. The Euclidean distances between glomerular activation patterns for component and blend responses at the OSN level showed that isoamyl acetate (a compound typical of banana to the human nose) was ranked highest among all components and suggest it as a key component of the banana blend (Figure 6A, Table 2). Just like the banana blend, isoamyl acetate has been shown to be a highly attractive component for *Drosophila* (Ayyub et al., 1990). Notably, in our analysis the majority of components became more similar to the blend representation at the PN level in comparison to the OSN representations. This result is substantiated by the fact that the average component-blend similarity was significantly higher in PNs compared to OSNs (Figure 6). This similarity change indicates that the functional representation of the individual banana components is modulated at the output level by the AL network. Such a processing mechanism might tune the olfactory system to categorize individual banana components with their naturally occurring odor source and to improve the fly's ability to detect suitable food sources against an environmental odor background. Further experiments are necessary to analyze whether the fly perceives the individual banana components as attractive as the banana blend itself.

Relevance of Glomerulus DM2 for Banana Perception

Glomerulus DM2 displayed the strongest responses both to single components and to the complete banana blend (Figure 3) which is in accordance with the study by Münch et al. (2013). This glomerulus was in addition the only glomerulus that showed significant mixture suppression at the PN level (Figure 5). We propose that this glomerulus has an important role in eliciting fly attraction to an attractive banana odor blend. Indeed, Semmelhack and Wang showed that innate fly behavior can be mediated by activity in individual glomeruli in the *Drosophila* AL (Semmelhack and Wang, 2009). Moreover, a recent study by Knaden et al. that analyzed the coding of odor valence in the*Drosophila* AL, clearly shows that glomerulus DM2 was significantly stronger activated by attractive odorants (Knaden et al., 2012). Future experiments using flies with an Or22a knock-out will shed further light on the behavioral relevance of this glomerulus regarding attractive natural odor sources.

The Impact of Natural Odor Concentrations using Functional GC-I

We combined the proven powers of two well established experimental designs, calcium imaging and GC fragmentation of a natural odor extract. The odor components were used as stimuli and presented to the fly while optophysiological measurements of the different processing levels in the AL were performed. In these functional GC-I experiments we were able to simultaneously investigate the majority of OSNs and PNs, respectively, during a single GC run.

To identify natural blend components which activate OSNs in insects common bioassays like GC-coupled electroantennographic detection (GC-EAD) (Arn and Rauscher, 1975; Struble and Arn, 1984) or GC-coupled single sensillum recordings (GC-SSR) (Stensmyr et al., 2003; Stökl et al., 2010) provided excellent data for odor responses in the periphery. While GC-EAD is a relative simple technique which allows conclusions about component activity in the whole insect antenna, GC-SSR experiments allow in addition the measurement of response profiles of specific sensillum types. The functional GC-I technique emerges as a significant expansion of these classical combined GC/electrophysiology techniques since it offers the investigation of olfactory processing in whole neuronal populations under near-natural conditions, meaning the sensory system can be tested under conditions where behavior is most relevant. It will thus be a highly useful tool in future investigations of insect-insect and insect-plant chemical interactions, and could be extended also to other animal groups.

AUTHOR CONTRIBUTIONS

Marco Schubert planned and carried out all the experiments; Silke Sachse and Bill S. Hansson together conceived and directed the project; Marco Schubert and Silke Sachse interpreted the results, prepared the figures, and wrote the paper.

ACKNOWLEDGMENTS

We are grateful to Ales Svatos for helping with the identification of banana components, to Mathias Ditzen for programming data analyzing tools in IDL, to Andreas Reinecke for helping in the development of the transfer line head and MS data analysis, to Veit Grabe for performing Amira™ reconstructions of the antennal lobe and to Silke Trautheim and Kerstin Weniger for excellent technical assistance. This study was supported by the Max Planck Society and by the Federal Ministry of Education and Research (BMBF research grant to Silke Sachse).

REFERENCES

1. Akers, R. P., and Getz, W. M. (1993). Response of olfactory receptor neurons in honeybees to odorants and their binary mixtures. *J. Comp. Physiol. A* 173, 169–185.

2. Arn, H., and Rauscher, S. (1975). The electroantennographic detector - A selective and sensitive tool in the gas chromatographic analysis of insect pheromones. *Z. Naturforschg.* 30c, 722–725.

3. Ayyub, C., Paranjape, J., Rodrigues, V., and Siddiqi, O. (1990). Genetics of olfactory behavior in *Drosophila melanogaster.J. Neurogenet.* 6, 243–262.

4. Bhandawat, V., Olsen, S. R., Gouwens, N. W., Schlief, M. L., and Wilson, R. I. (2007). Sensory processing in the Drosophila antennal lobe increases reliability and separability of ensemble odor representations. *Nat. Neurosci.* 10, 1474–1482. doi: 10.1038/nn1976.

5. Brand, A. H., and Perrimon, N. (1993). Targeted gene expression as a means of altering cell fates and generating dominant phenotypes. *Development* 118, 401–415.

6. Brown, S. L., Joseph, J., and Stopfer, M. (2005). Encoding a temporally strutured stimulus with a temporally structured neural representation. *Nat. Neurosci.* 8, 1568–1576. doi: 10.1038/nn1559

7. Carlsson, M. A., Chong, K. Y., Daniels, W., Hansson, B. S., and Pearce, T. C. (2007). Component information is preserved in glomerular responses to binary odor mixtures in the moth

Spodoptera littoralis. Chem. Senses 32, 433–443. doi: 10.1093/ chemse/bjm009

8. Chou, Y.-H., Spletter, M. L., Yaksi, E., Leong, J. C. S., Wilson, R. I., and Luo, L. (2010). Diversity and wiring variability of olfactory local interneurons in the Drosophila antennal lobe. *Nat. Neurosci.* 13, 439–449. doi: 10.1038/nn.2489

9. Couto, A., Alenius, M., and Dickson, B. J. (2005). Molecular, anatomical, and functional organization of the Drosophila olfactory system. *Curr. Biol.* 15, 1535–1547. doi: 10.1016/j. cub.2005.07.034

10. De Bruyne, M., Foster, K., and Carlson, J. R. (2001). Odor coding in the *Drosophila antenna. Neuron* 30, 537–552. doi: 10.1016/ S0896-6273(01)00289-6

11. Deisig, N., Giurfa, M., Lachnit, H., and Sandoz, J.-C. (2006). Neural representation of olfactory mixtures in the honeybee antennal lobe. *Eur. J. Neurosci.* 24, 1161–1174. doi: 10.1111/j.1460-9568.2006.04959.x

12. Deisig, N., Giurfa, M., and Sandoz, J. C. (2010). Antennal lobe processing increases separability of odor mixture representations in the honeybee. *J. Neurophysiol.* 103, 2185–2194. doi: 10.1152/ jn.00342.2009

13. Duchamp-Viret, P., Duchamp, A., and Chaput, M. A. (2003). Single olfactory sensory neurons simultaneously integrate the components of an odour mixture. *Eur. J. Neurosci.* 18, 2690–2696. doi: 10.1111/j.1460-9568.2003.03001.x

14. Fernandez, P. C., Locatelli, F. F., Person-Rennell, N., Deleo, G., and Smith, B. H. (2009). Associative conditioning tunes transient dynamics of early olfactory processing. *J. Neurosci.* 29, 10191–10202. doi: 10.1523/jneurosci.1874-09.2009

15. Fiala, A., Spall, T., Diegelmann, S., Eisermann, B., Sachse, S., Devaud, J. M., et al. (2002). Genetically expressed cameleon in *Drosophila melanogaster* is used to visualize olfactory information in projection neurons. *Curr. Biol.* 12, 1877–1884. doi: 10.1016/ S0960-9822(02)01239-3

16. Fishilevich, E., and Vosshall, L. B. (2005). Genetic and functional subdivision of the Drosophila antennal lobe. *Curr. Biol.* 15, 1548–1553. doi: 10.1016/j.cub.2005.07.066

17. Galizia, C. G., and Sachse, S. (2010). "Odor coding in insects," in *The Neurobiology of Olfaction*, ed A. Menini (Boca Raton; London; New York: CRC Press), 35–70.

18. Gao, Q., Yuan, B., and Chess, A. (2000). Convergent projections of Drosophila olfactory neurons to specific glomeruli in the antennal lobe. *Nat. Neurosci.* 3, 780–785. doi: 10.1038/77680

19. Grossman, K. J., Mallik, A. K., Ross, J., Kay, L. M., and Issa, N. P. (2008). Glomerular activation patterns and the perception of odor mixtures. *Eur. J. Neurosci.* 27, 2676–2685. doi: 10.1111/j.1460-9568.2008.06213.x

20. Guerrieri, F., Schubert, M., Sandoz, J. C., and Giurfa, M. (2005). Perceptual and neural olfactory similarity in honeybees. *PLoS Biol.* 3:e60. doi: 10.1371/journal.pbio.0030060

21. Hallem, E. A., and Carlson, J. R. (2006). Coding of odors by a receptor repertoire. *Cell* 125, 143–160. doi: 10.1016/j.cell.2006.01.050

22. Hansson, B. S., Knaden, M., Sachse, S., Stensmyr, M. C., and Wicher, D. (2010). Towards plant-odor-related olfactory neuroethology in Drosophila. *Chemoecology* 20, 51–61. doi: 10.1007/s00049-009-0033-7

23. Jordán, M. J. (2001). Aromatic profile of aqueous banana essence and banana fruit by Gas Chromatography-Mass Spectrometry (GC-MS) and Gas Chromatography-Olfactometry (GC-O). *J. Agr. Food Chem.* 49, 4813–4817. doi: 10.1021/jf010471k

24. Knaden, M., Strutz, A., Ahsan, J., Sachse, S., and Hansson, B. S. (2012). Spatial representation of odorant valence in an insect brain. *Cell Rep.* 1, 392–399. doi: 10.1016/j.celrep.2012.03.002

25. Kuebler, L. S., Olsson, S. B., Weniger, R., and Hansson, B. S. (2011). Neuronal processing of complex mixtures establishes a unique odor representation in the moth antennal lobe. *Front. Neural Cir.* 5:7. doi: 10.3389/fncir.2011.00007

26. Lachaise, D., and Silvain, J.-F. (2004). How two Afrotropical endemics made two cosmopolitan human commensals: the *Drosophila melanogaster-D. simulans* palaeogeographic riddle. *Genetica* 120, 17–39. doi: 10.1023/B:GENE.0000017627.27537.ef

27. Laissue, P. P., Reiter, C., Hiesinger, P. R., Halter, S., Fischbach, K. F., and Stocker, R. F. (1999). Three-dimensional reconstruction of the antennal lobe in *Drosophila melanogaster*. *J. Comp. Neurol.* 405, 543–552. doi: 10.1002/(SICI)1096-9861(19990322)405:4

28. Larsson, M. C., Domingos, A. I., Jones, W. D., Chiappe, M. E., Amrein, H., and Vosshall, L. B. (2004). Or83b encodes a broadly expressed odorant receptor essential for Drosophila olfaction. *Neuron* 43, 703–714. doi: 10.1016/j.neuron.2004.08.019

29. Lin, D. Y., Shea, S. D., and Katz, L. C. (2006). Representation of natural stimuli in the rodent main olfactory bulb. *Neuron* 50, 937–949. doi: 10.1016/j.neuron.2006.03.021

30. Münch, D., Schmeichel, B., Silbering, A. F., and Galizia, C. G. (2013). Weaker ligands can dominate an odor blend due to syntopic interactions. *Chem. Senses* 38, 293–304. doi: 10.1093/chemse/bjs138

31. Nakai, J., Ohkura, M., and Imoto, K. (2001). A high signal-to-noise Ca(2+) probe composed of a single green fluorescent protein. *Nat. Biotechnol.* 19, 137–141. doi: 10.1038/84397

32. Ng, M., Roorda, R. D., Lima, S. Q., Zemelman, B. V., Morcillo, P., and Miesenbock, G. (2002). Transmission of olfactory information between three populations of neurons in the antennal lobe of the fly. *Neuron* 36, 463–474. doi: 10.1016/S0896-6273(02)00975-3

33. Okada, R., Awasaki, T., and Ito, K. (2009). Gamma-aminobutyric acid (GABA)-mediated neural connections in the Drosophila antennal lobe. *J. Comp. Neurol.* 514, 74–91. doi: 10.1002/cne.21971

34. Olsen, S. R., Bhandawat, V., and Wilson, R. I. (2010). Divisive normalization in olfactory population codes. *Neuron* 66, 287–299. doi: 10.1016/j.neuron.2010.04.009

35. Pelz, D., Roeske, T., Syed, Z., Bruyne, M. D., and Galizia, C. G. (2006). The molecular receptive range of an olfactory receptor *in vivo* (*Drosophila melanogaster* Or22a). *J. Neurobiol.* 66, 1544–1563. doi: 10.1002/neu.20333

36. Rospars, J.-P., Lansky, P., Chaput, M., and Duchamp-Viret, P. (2008). Competitive and noncompetitive odorant interactions in the early neural coding of odorant mixtures. *J. Neurosci.* 28, 2659–2666. doi: 10.1523/jneurosci.4670-07.2008

37. Sachse, S., Rappert, A., and Galizia, C. G. (1999). The spatial representation of chemical structures in the antennal lobes of honeybees: steps towards the olfactory code. *Eur. J. Neurosci.* 11, 3970–3982. doi: 10.1046/j.1460-9568.1999.00826.x

38. Seki, Y., Rybak, J., Wicher, D., Sachse, S., and Hansson, B. S. (2010). Physiological and morphological characterization of local interneurons in the Drosophila antennal lobe. *J. Neurophysiol.* 104, 1007–1019. doi: 10.1152/jn.00249.2010

39. Semmelhack, J. L., and Wang, J. W. (2009). Select Drosophila glomeruli mediate innate olfactory attraction and aversion. *Nature* 459, 218–223. doi: 10.1038/nature07983

40. Shang, Y., Claridge-Chang, A., Sjulson, L., Pypaert, M., and Miesenböck, G. (2007). Excitatory local circuits and their implications for olfactory processing in the fly antennal lobe. *Cell* 128, 601–612. doi: 10.1016/j.cell.2006.12.034

41. Shiota, H. (1993). New esteric components in the volatiles of banana fruit (*Musa sapientum* L.). *J. Agric. Food Chem.* 41, 2056–2062. doi: 10.1021/jf00035a046

42. Silbering, A. F., and Galizia, C. G. (2007). Processing of odor mixtures in the Drosophila antennal lobe reveals both global inhibition and glomerulus-specific interactions. *J. Neurosci.* 27, 11966–11977. doi: 10.1523/jneurosci.3099-07.2007

43. Silbering, A. F., Okada, R., Ito, K., and Galizia, C. G. (2008). Olfactory information processing in the Drosophila antennal lobe: anything goes? *J. Neurosci.* 28, 13075–13087. doi: 10.1523/jneurosci.2973-08.2008

44. Silbering, A. F., Rytz, R., Grosjean, Y., Abuin, L., Ramdya, P., Jefferis, G. S. X. E., et al. (2011). Complementary function and integrated wiring of the evolutionarily distinct Drosophila olfactory subsystems. *J. Neurosci.* 31, 13357–13375. doi: 10.1523/jneurosci.2360-11.2011

45. Stensmyr, M. C., Dweck, H. K. M., Farhan, A., Ibba, I., Strutz, A., Mukunda, L., et al. (2012). A conserved dedicated olfactory circuit for detecting harmful microbes in Drosophila. *Cell* 151, 1345–1357. doi: 10.1016/j.cell.2012.09.046

46. Stensmyr, M. C., Giordano, E., Balloi, A., Angioy, A. M., and Hansson, B. S. (2003). Novel natural ligands for Drosophila

olfactory receptor neurones. *J. Exp. Biol.* 206, 715–724. doi: 10.1242/jeb.00143

47. Stocker, R. F., Heimbeck, G., Gendre, N., and De Belle, J. S. (1997). Neuroblast ablation in Drosophila P[GAL4] lines reveals origins of olfactory interneurons. *J. Neurobiol.* 32, 443–456. doi: 10.1002/(SICI)1097-4695(199705)32:5

48. Stökl, J., Strutz, A., Dafni, A., Svatos, A., Doubsky, J., Knaden, M., et al. (2010). A deceptive pollination system targeting drosophilids through olfactory mimicry of yeast. *Curr. Biol.* 20, 1846–1852. doi: 10.1016/j.cub.2010.09.033

49. Struble, D. L., and Arn, H. (1984). "Combined gas chromatography and electroantennogramm recording of insect olfactory responses," in *Techniques in Pheromone Research*, ed H. E. Hummel (New York, NY: Springer-Verlag), 161–177.

50. Strutz, A., Voeller, T., Riemensperger, T., Fiala, A., and Sachse, S. (2012). "Calcium imaging of neural activity in the olfactory system of Drosophila," in *Genetically Encoded Functional Indicators*, ed J.-R. Martin (New York, NY: Springer Science+Business Media LLC), 43–70. doi: 10.1007/978-1-62703-014-4_3

51. Sturtevant, A. H. (1921). *The North American species of Drosophila / by A.H. Sturtevant*. Washington, DC: Carnegie Institution of Washington.

52. Tabor, R., Yaksi, E., Weislogel, J. M., and Friedrich, R. W. (2004). Processing of odor mixtures in the zebrafish olfactory bulb. *J. Neurosci.* 24, 6611–6620. doi: 10.1523/jneurosci.1834-04.2004

53. Vosshall, L. B., and Stocker, R. F. (2007). Molecular architecture of smell and taste in Drosophila. *Annu. Rev. Neurosci.* 30, 505–533. doi: 10.1146/annurev.neuro.30.051606.094306

54. Vosshall, L. B., Wong, A. M., and Axel, R. (2000). An olfactory sensory map in the fly brain. *Cell* 102, 147–159. doi: 10.1016/S0092-8674(00)00021-0

55. Wang, J. W., Wong, A. M., Flores, J., Vosshall, L. B., and Axel, R. (2003). Two-photon calcium imaging reveals an odor-evoked map of activity in the fly brain. *Cell* 112, 271–282. doi: 10.1016/S0092-8674(03)00004-7

56. Wilson, R. I. (2011). Understanding the functional consequences of synaptic specialization: insight from the Drosophila antennal

lobe. *Curr. Opin. Neurobiol.* 21, 254–260. doi: 10.1016/j. conb.2011.03.002

57. Wilson, R. I., and Laurent, G. (2005). Role of GABAergic inhibition in shaping odor-evoked spatiotemporal patterns in the Drosophila antennal lobe. *J. Neurosci.* 25, 9069–9079. doi: 10.1523/JNEUROSCI.2070-05.2005

Effect of Bicarbonate on the Mineralization of Methyldiethanolamine by using UV/H2O2

Sabtanti Harimurti[1, 2], Anisa Ur Rahmah[1],
Abdul A. Omar[1], and Thanapalan Murugesan[1]

[1]Department of Chemical Engineering, Universiti Teknologi PETRONAS, Bandar Seri Iskandar, Tronoh, 31750, Perak Darul Ridzwan, Malaysia

[2]Department of Pharmacy, Faculty of Medicine and Health Science, University Muhammadiyah Yogyakarta, Yogyakarta, 55183, Indonesia

ABSTRACT

The presence of bicarbonate affects the degradation efficiency of effluents containing aqueous methyldiethanolamine (MDEA) solution leaving the CO_2 absorption/regeneration unit of natural gas processing units. In the present study the effect of bicarbonate at three different pH conditions of (acidic, neutral and alkaline) simulated MDEA solution were conducted, by the addition of six different concentration of $NaHCO_3$ (0.025, 0.05, 0.075, 0.1, 0.125 and 0.15 M). The presence of bicarbonate increased the mineralization of MDEA when the reaction was conducted at neutral initial pH conditions, where as the MDEA mineralization was reduced when the reaction was conducted at alkaline pH condition.

INTRODUCTION

Methyldiethanolamine (MDEA) is one of the alkanolamines that is commonly used for the removal of acidic gases (such as H_2S and CO_2) from natural gas (Kohl and Nielsen, 1997). Removal of acidic gas from the natural gas is necessary since the acidic gases cause corrosion in pipeline and processing equipment, also reduce the heating value which has an effect on the price of natural gas. MDEA has two ethanol functional groups and one methyl group attached to a nitrogen atom. Due to the existence of nitrogen atom with a pair of free electrons, MDEA forms weak base with water, hence MDEA is often used for scrubbing/sweetening of acidic gases (CO_2 and H_2S) from raw natural gas. Aqueous MDEA solution chemically binds with the acidic gases and when heated it releases the absorbed gases (Kohl and Nielsen, 1997).

During shutdown and maintenance of the processing equipments, high concentrations of residual MDEA will be carried over into the effluent. Conventional biological oxidation process is not effective for the treatment of effluent containing MDEA. Furhacker et al. (2003) reported that MDEA was not biodegradable in the bioreactor for a test period of 28 days. Therefore, an Advanced Oxidation Process (AOP) i.e., UV/H_2O_2 was used to treat the aqueous MDEA solution, by which approximately 86 % of TOC was removed (Harimurti et al., 2012). During the absorption/scrubbing process, the bicarbonate is also generated and present along with MDEA in the effluent stream. A number of researchers have reported that the presence of bicarbonate during the AOP's reduces the degradation efficiency. Mehrvar et al. (2001) reported that the degradation rate of tetrahydrofuran (THF) and 1,4-dioxane (DIOX) were affected/reduced due to the presence of bicarbonate in the system. During the UV/H_2O_2 process, the presence of bicarbonate was reported to be strongly inhibit the degradation of organophosphorus pesticides namely, malathian and diazinon (Fadaei et al., 2012) The addition of inorganic ion such as bicarbonate in the Procion H-exl dyes solution gave an adverse effect on the decolorization rate of dye using Fenton's process, as reported by Riga et al. (2007). Daneshvar et al. (2007) concluded that the presence of bicarbonate during the photooxidative degradation (UV/H$_2$O$_2$) reduced the degradation rate of 4-nitrophenol (4-NP) (Daneshvaret al., 2007). Muruganandham and Swaminathan (2004) studied the effect of bicarbonate during the photodecolorization of reactive azo dye (Reactive orange 4) and concluded that only 3.58% of decolorization was achieved (Muruganandham and Swaminathan, 2004). Klamerth et al. (2010) reported that bicarbonate competes with the organic contaminant for hydroxyl radical during the degradation of municipal wastewater using photo-Fenton and hence the presence of bicarbonate reduced the degradation efficiency of the effluent. Degradation of Atrazine by manganese-catalysed ozonation was also inhibited by the presence of bicarbonate (Ma and Graham, 2000). Based on the fore going observation from the available literature, the present study will focus on the effect of bicarbonate on the mineralization of effluents containing MDEA using UV/H$_2$O$_2$ process.

MATERIALS AND METHODS

Methyldiethanolamine (MDEA), potassium permangate (KMnO$_4$), sulfuric acid (H$_2$SO$_4$) and hydrogen peroxide (H$_2$O$_2$) were obtained from Merck (Germany). Sodium hydroxide (NaOH) and sodium bicarbonate (NaHCO$_3$) were obtained from RM Chemicals (Malaysia). Simulated MDEA solution was prepared by dissolving a desired amount of MDEA in distilled water.

All the experiments were conducted in 700 mL cylindrical stirred jacketed glass reactor to monitor the progress of mineralization. The photoreactor was equipped with 8 Watt low pressure Hg vapor lamp GPH295T5L (which produces UV light at 254 nm was made in USA with serial no. EC90277), a current-voltage control unit and an opening at the top for sample collection Intensity of UV lamp was measured by using UV radiometer (Cole-Parmer model: 97651-10 with sensor UV 254 nm model: 97651-20). The pH value of the solution was measured using pH meter (HACH-senion1) and the adjustment was carried out using NaOH or H$_2$SO$_4$ accordingly. The temperature of reaction was adjusted/maintained by circulating cooling water through the jacket.

To study the effect of bicarbonate on the degradation of MDEA, $NaHCO_3$ with known concentrations were added into the mixture and allowed to dissolve before the start of experiments. Liquid samples were withdrawn at specific time intervals and the TOC of the samples were measured using TOC analyzer (Shimadzu TOC-V$_{CSH}$). H_2O_2 concentration in the solution during the photochemical oxidation process was monitored by titrating the samples using standard $KMnO_4$ solution (Mendham et al., 2000).

RESULTS AND DISCUSSION

The present research includes the preliminary studies on the individual effect of UV and H_2O_2 as well as the combination of UV/H_2O_2 on the photochemical mineralization of aqueous MDEA solution and later extended to study the effect of the presence of bicarbonate on the photochemical oxidation system. The experiments were conducted using the following optimum conditions: Intensity of UV lamp = 12.06 mW/cm^2, irradiation time = 3 hours, oxidation temperature = 30°C, [MDEA]$_0$ = 2000 ppm, [H_2O_2]$_0$ = 0.22 M (Behnajady et al., 2008).

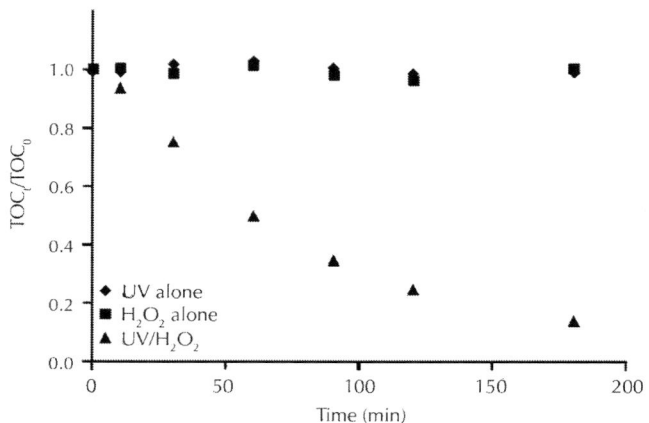

Figure 1: Individual effect of UV, H_2O_2 and the combination of UV/H_2O_2 on the MDEA mineralization.

To study the bicarbonate effect, six different concentrations of $NaHCO_3$ (0.025, 0.05, 0.075, 0.1, 0,125, and 0.15 M) were used.

The initial pH of reaction was = 7, since at the acidic conditions, the bicarbonate will be neutralized to form CO_2 and hence for the present work two different initial pH conditions (7 and 10.18) were used.

UV light and H_2O_2 process are well-known for the degradation of many organic compounds in aqueous solution. The capability of UV light to degrade the organic compound follows photolysis mechanism. The organic compound absorbs UV spectrum and then results in an excited (organic) compound which later decomposed to form a product (Massachelein, 2002; Oppenlander, 2003; Lester et al., 2010; Seraghni et al., 2012). Based on the preliminary experiments it was observed that, UV spectrum at 254 nm (used in the present experiments) was not capable to remove the total organic carbon from the system (Fig. 1). The reason could be attributed to the fact that MDEA did not absorb the UV light at 254 nm, since the spectrum absorbed by MDEA was at 200 nm region, therefore, the direct photolysis did not occur (Harimurti et al., 2013). The capability of H_2O_2 to degrade organic compound is mainly due to the high reduction potential of H_2O_2 i.e., +1.8 V. This reduction potential indicates the high tendency of H_2O_2 to act as an oxidant which refers to direct electron-transfer reaction between organic compound and H_2O_2 (Petri et al., 2011). The results of the present experiments showed no degradation when the H_2O_2 alone was used, indicating that H_2O_2 alone was not capable to remove the total organic carbon in the aqueous MDEA solution (Fig. 1). This might be due to the reduction potential of H_2O_2 which is not sufficient for the oxidation process. The photolysis resistance of MDEA toward UV light and H_2O_2 was in agreement with the observation of (Xu et al., 2009), based on their studies on the photolysis resistance of dimethyl phthalate against UV photolysis and H_2O_2. However, the reduction of total organic carbon was found when the UV and H_2O_2 were applied in combination. The total organic carbon was reduced to a certain level (Fig. 1) which was due to the hydroxyl radical generated from H_2O_2 photolysis. It is well known that H_2O_2 strongly absorbs UV spectrum at 254 nm (Massachelein, 2002). Therefore the probability of H_2O_2 photolysis to generate hydroxyl radicals is very high. In other words, the combination of UV and H_2O_2 will generate hydroxyl radical which plays an important role during the degradation of many recalcitrant organic contaminants (Daneshvar et al., 2007; Lester et al., 2010; Behnajady et al., 2008; Abramovic et al., 2010).

Absorption/scrubbing of CO_2 by aqueous MDEA solution occurs according to the following reactions:

Ionization of water:

$$H_2O \rightarrow H^+ + OH^-$$

(1)

Hydrolysis and ionization of dissolve CO_2:

$$CO_2 + H_2O \rightarrow HCO_3^- + H^+$$

(2)

Protonation of MDEA:

$$R_2NCH_3 + H^+ \rightarrow R_2NCH_4^+$$

(3)

Acid-basic reaction with the amine:

$$R_2NCH^+ + HCO_3^- \leftrightarrow R_2NCH_3 + H_2O + CO_2$$

(4)

During the scheduled shut down of scrubbing unit in the natural gas plant, bicarbonate (HCO_3^-) is expected to present in the effluents leaving the gas processing unit. Generally, the presence of bicarbonate in the AOP's will act as a scavenger for hydroxyl radical. Bicarbonate (HCO_3^-) reacts with hydroxyl radical ($HO\bullet$) to form bicarbonate radical ($HCO_3\bullet$). This radical is also a well-known oxidant, but much less reactive compared to hydroxyl radical (Riga et al., 2007; Daneshvar et al., 2007; Jones, 1999; Andreozzi et al., 1999; Chiang et al., 2006). Consequently, the degree of oxidation is expected to be less.

In order to study the effect of bicarbonate on the mineralization of aqueous MDEA solution using the combination of UV/H_2O_2 process, experiments were conducted at two different initial pH (7 and 10.18) conditions and six different concentrations of $NaHCO_3$ in the aqueous MDEA solution.

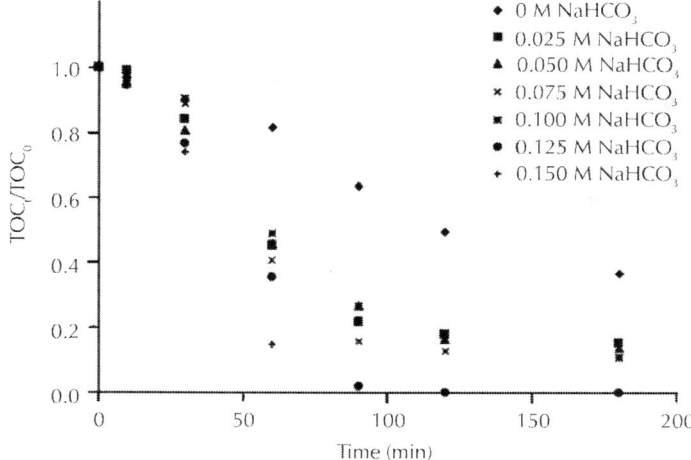

Figure 2: Total organic carbon profile during the degradation of MDEA in the presence of $NaHCO_3$ using UV/H_2O_2, Inital pH = 7.

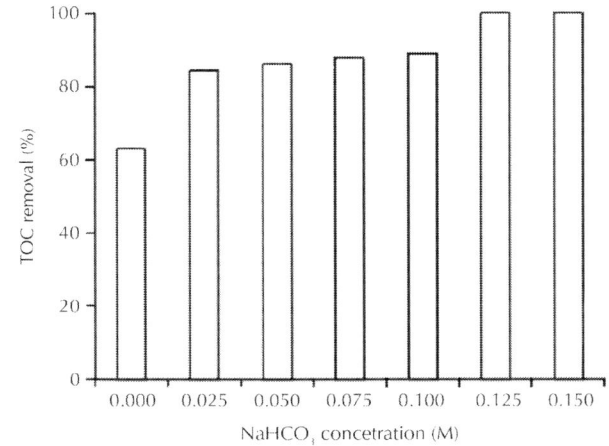

Figure 3: Percentage TOC removal achieved at initial pH reaction = 7.

At acidic pH conditions (pH <7) the bicarbonate will get neutralized (Eq. 5) and the product i.e., CO_2 will be released from the system, hence there will be no effect on the mineralization process.

$$HCO_2^- + H^+ \to H_2O + CO_2$$

(5)

When the initial pH of the process was approximately 7, the presence of bicarbonate in the synthetic aq. MDEA solution increased the mineralization of MDEA (Fig. 2). The degree of MDEA mineralization was increased by increasing of $NaHCO_3$ concentration in the system and complete mineralization was achieved when the concentration of $NaHCO_3$ was = 0.125 M (Fig. 3). This trend can be explained as: The capability of bicarbonate that can act as a good buffer was found at this condition. The pH during the mineralization process was maintained at 7 (Fig. 4). Bicarbonate is an amphoteric ion that can act either as an acid which can donate its H+ to form CO_3^{2-} or as a base which is capable to accept an H+ to form H_2CO_3. Formation of organic acid during the degradation process, reduce the pH of the system. At this pH condition (pH = 7), the bicarbonate reacts with the hydroxyl radical (which is available in the system) to form bicarbonate radical, however, the formation rate of bicarbonate radical is less (8.5x106 M-1 s-1). Equation 6-7 shows the formation of bicarbonate radical in the system (Oppenlander, 2003; Andreozzi et al., 1999; Tang, 2003).

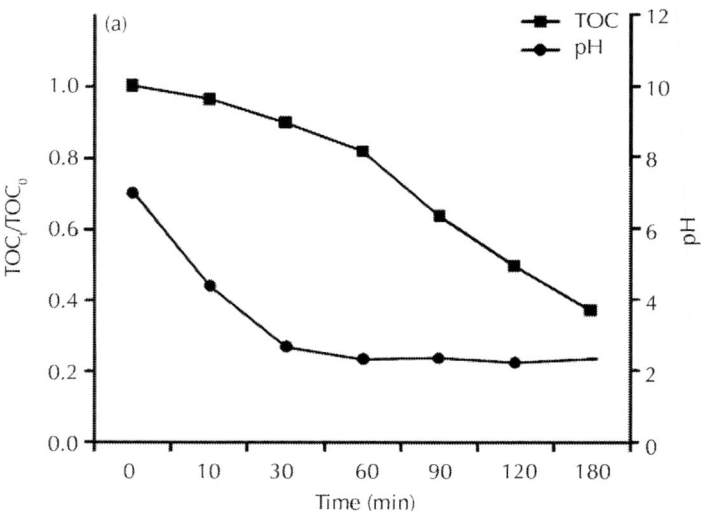

(a)

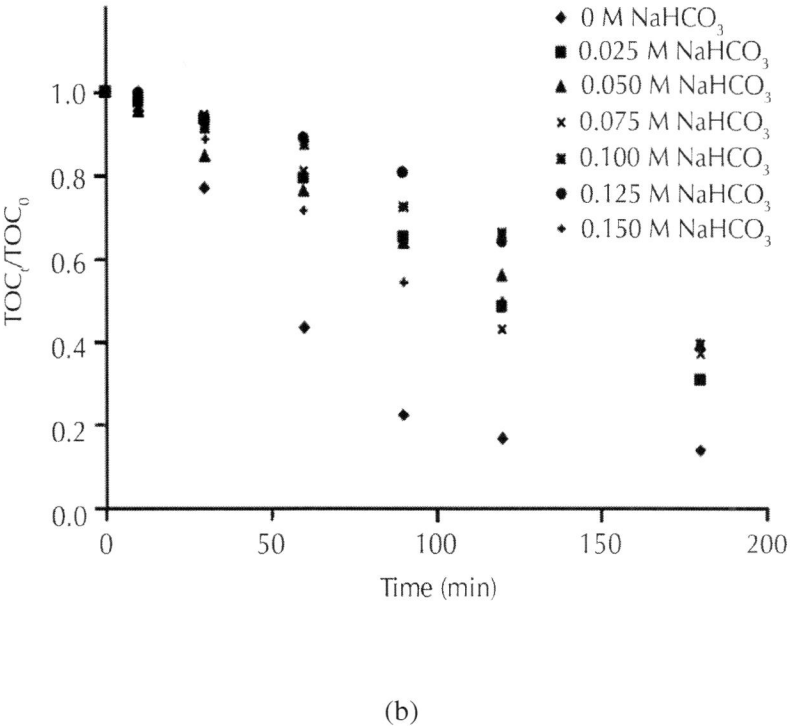

(b)

Figure 4: pH profile during the degradation of MDEA using UV/H$_2$O$_2$ at initial pH = 7

$$H_2O_2 + hv \rightarrow HO\bullet \tag{6}$$

$$HO\bullet + HCO_3^- \rightarrow HO^- + HCO_3\bullet \tag{7}$$

Klare et al., (2000) reported that free electron pair of nitrogen atom of amine compounds are in un-protonated form for pH = 7 and under these conditions, more active sites for oxidation by hydroxyl radical are available.

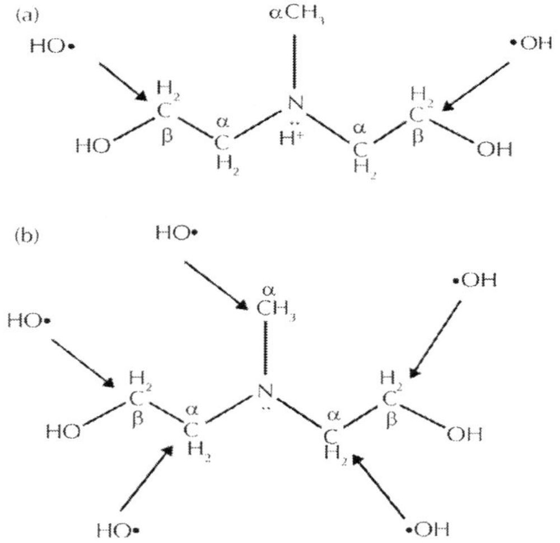

Figure 5: Active sites of MDEA at (a) Acidic and (b) High pH (pH≥7) for the oxidation by hydroxyl radical.

Thus, the presence of HCO_3^- in the solution maintained the free electron pair of nitrogen atom of MDEA in un-protonated condition and hence more active sites for reaction are always provided (Fig. 5) which in turn leads to higher degradation of MDEA.

The effect of the presence of bicarbonate during mineralization of aq. MDEA solution was also studied at alkaline pH conditions. Based on the preliminary studies, the optimum pH for mineralization process of aqueous MDEA solution was found to be 10.18, hence this pH was chosen for the present study at alkaline pH condition (Harimurti et al., 2012). The presence of bicarbonate (HCO_3^-) during the UV/H_2O_2 mineralization of simulated aq. MDEA solution at an initial pH of 10.18 was found to reduce the mineralization efficiency (Fig. 6). Approximately 25% of TOC removal was reduced when the concentration of $NaHCO_3$ was = 0.050 M (Fig. 7). This might be due to the presence of bicarbonate in the system which act as a good buffer and maintain the pH at constant and not allowed to drop to lower pH levels (Fig. 8). At alkaline pH condition, the bicarbonate (HCO_3^-) was converted into carbonate (CO_3^{2-}) and then react with hydroxyl radical to form carbonate radical. Reaction between carbonate and hydroxyl radical is shown below:

$$HO\bullet + CO_3^{2-} \rightarrow HO^- + CO_3^{-}\bullet$$

(8)

Since the reaction rate of carbonate radical formation is high (i.e., 3.9×10^8 M-1 s-1) and the scavenger reaction was significant enough to reduce the concentration of hydroxyl radical in the system that act as an important oxidant in the UV/H_2O_2 process, hence the reduction in MDEA mineralization.

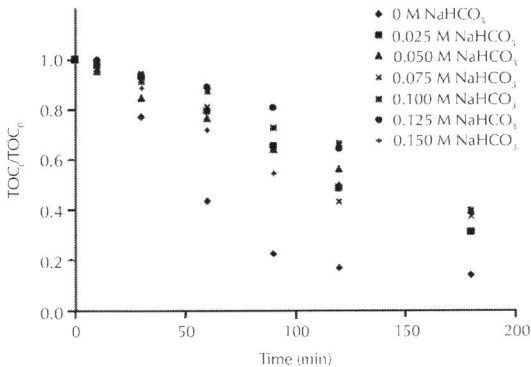

Figure 6: Total organic carbon profile during the degradation of MDEA in the presence of NaHCO$_3$ using UV/H_2O_2 at initial pH = 10.18.

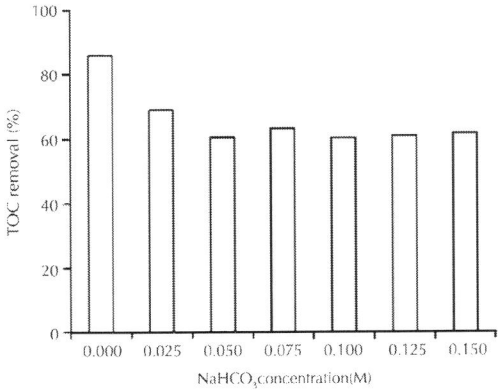

Figure 7: Percentage TOC removal achieved at initial pH reaction = 10.18.

Moreover, H_2O_2 tend to ionize to form hydroperoxide anion (HO_2^-) at high pH with pKa equals to 11.6 (Xu et al., 2009; Ren et al., 2010). Hydroperoxide anion is well-known to be a strong scavenger to hydroxyl radical (Eq. 9 and 10).

$$H_2O_2 \leftrightarrow HO_2^- + H^+$$

(9)

$$HO\bullet + HO_2^- \rightarrow HO_2\bullet + HO^-$$

(10)

Reaction between hydroperoxide anion (HO_2^-) and hydroxyl radical ($HO\bullet$) generates a less reactive radical species i.e., hydroperoxyl radical ($HO_2\bullet$) which is very effective to reduce the hydroxyl radical in the system and hence, the mineralization efficiency decreased. Even though bicarbonate reacts with the hydroxyl radical to form bicarbonate radical in the neutral pH (initial pH $\approx$7), the presence of bicarbonate radical did not interfere in the photodegradation process, since the reaction rate of bicarbonate and hydroxyl radical was less. Based on these analysis, it can be concluded that at neutral pH, due to the enhancement of active sites of MDEA for oxidation the bicarbonate increases the TOC removal. However, at alkaline pH condition of reaction, the bicarbonate acts as a strong scavenger to hydroxyl radical by converting to carbonate which further reacts with hydroxyl radical and reduce the TOC removal.

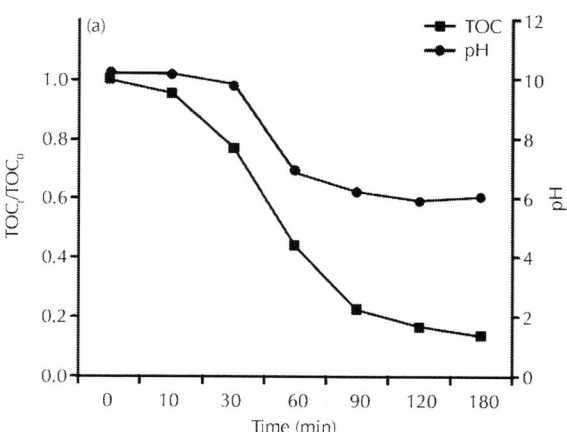

(a)

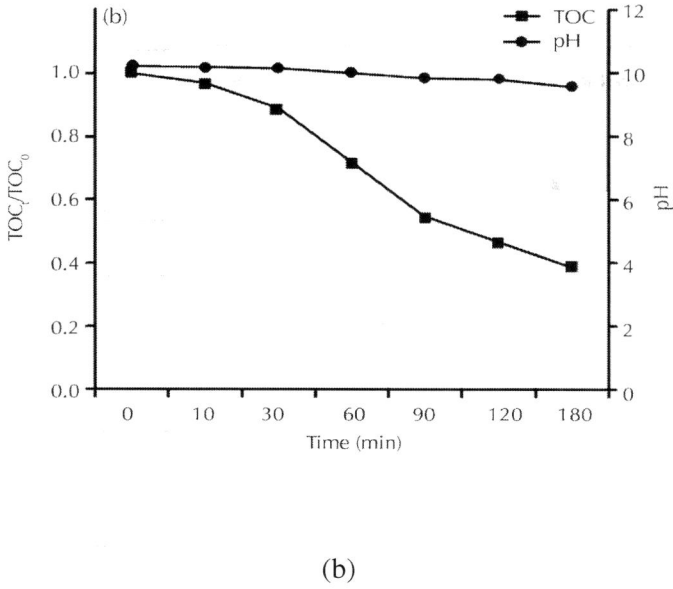

(b)

Figure 8: TOC and pH profile during the degradation of MDEA using UV/ H_2O_2 at initial pH = 10.18 at (a) 0 M NaHCO$_3$ and (b) 0.15 M NaHCO$_3$

CONCLUSIONS

Based on the present experiments it can be concluded that the presence of bicarbonate increase the mineralization rate of aqueous MDEA solution when the reaction was conducted at neutral pH conditions (pH = 7) but slows down the mineralization rate of MDEA when the reaction was conducted at pH = 10.18. The enhancement of mineralization efficiency is attributed to the capability of bicarbonate to maintain the pH at 7, during the reaction and hence the active sites for oxidation by hydroxyl radical are always available. This information will be of useful for the design and scale up of the UV/H_2O_2 oxidation process for the photochemical degradation of effluents from CO_2 absorber/scrubbing units which normally contains bicarbonate along with MDEA.

ACKNOWLEDGMENTS

The scholarship to Sabtanti Harimurti, under the Graduate Assistant Scheme from Universiti Teknologi PETRONAS, is highly acknowledged.

REFERENCES

1.   Abramovic, B.F., N.D. Banic and D.V. Sojic, 2010. Degradation of thiacloprid in aqueous solution by UV and UV/H$_2$O$_2$ treatments. Chemosphere, 81: 114-119.

2.   Andreozzi, R., V. Caprio, A. Insola and R. Marotta, 1999. Advanced Oxidation Processes (AOP) for water purification and recovery. Catal. Today, 53: 51-59.

3.   Behnajady, M.A., N. Modirshahla, M. Shokri and B. Vahid, 2008. Investigation of the effect of ultrasonic waves on the enhancement of efficiency of direct photolysis and photooxidation processes on the removal of a model contaminant from textile industry. Global NEST J., 10: 8-15.

4.   Chiang, Y.P., Y.Y. Liang, C.N. Chang and A.C. Chao, 2006. Differentiating ozone direct and indirect reactions on decomposition of humic substances. Chemosphere, 65: 2395-2400.

5.   Daneshvar, N., M.A. Behnajady and Y.Z. Asghar, 2007. Photooxidative degradation of 4-nitrophenol (4-NP) in UV/H$_2$O$_2$ process: Influence of operational parameters and reaction mechanism. J. Hazard. Mater., 139: 275-279.

6.   Fadaei, A.M., M.H. Dehghani, A.H. Mahvi, S. Nasseri, N. Rastkari and M. Shayeghi, 2012. Degradation of organophosphorus pesticides in water during UV/H$_2$O$_2$ treatment: Role of sulphate and bicarbonate ions. E-J. Chem., 9: 2015-2022.

7.   Furhacker, M., A. Pressl and R. Allabashi, 2003. Aerobic biodegradability of methyldiethanolamine (MDEA) used in natural gas sweetening plants in batch tests and continuous flow experiments. Chemosphere, 52: 1743-1748.

8.   Harimurti, S., A.U. Rahmah , A.A. Omar and T. Murugesan, 2012. Application of response surface method in the degradation

of wastewater containing MDEA using UV/H_2O_2 advanced oxidation process. J. Applied Sci., 12: 1093-1099.

9. Harimurti, S., A.U. Rahmah, A.A. Omar and M. Thanapalan, 2013. Kinetics of methyldiethanolamine mineralization by using UV/H_2O_2 process. CLEAN-Soil Air Water, 41: 1165-1174.

10. Jones, C.W., 1999. Application of Hydrogen Peroxide and Derivatives. Royal Society of Chemistry, London, UK., Pages: 264.

11. Klamerth, N., L. Rizzo, S. Malato, M.I. Maldonado, A. Aguera and A.R. Fernandez-Alba, 2010. Degradation of fifteen emerging contaminants at µgL-1 initial concentrations by mild solar photo-Fenton in MWTP effluents. Water Res., 44: 545-554.

12. Klare, M., J. Scheen, K. Vogelsang, H. Jacobs and J.A. Broekaert, 2000. Degradation of short-chain alkyl- and alkanolamines by TiO_2- and Pt/TiO_2-assisted photocatalysis. Chemosphere, 41: 353-362.

13. Kohl, A. and R. Nielsen, 1997. Gas Purification. 5th Edn., Gulf Publishing Company, Houston, TX., USA.

14. Lester, Y., D. Avisar and H. Mamane, 2010. Photodegradation of the antibiotic sulphamethoxazole in water with UV/H_2O_2 advanced oxidation process. Environ. Technol., 31: 175-183.

15. Ma, J. and N.J. Graham, 2000. Degradation of atrazine by manganese-catalysed ozonation-influence of radical scavengers. Water Res., 34: 3822-3828.

16. Massachelein, W.J., 2002. Ultraviolet Light in Water and Wastewater Sanitation. Lewis Publishers, New York, USA.

17. Mehrvar, M., W.A. Anderson and M. Moo-Young, 2001. Photocatalytic degradation of aqueous organic solvents in the presence of hydroxyl radical scavengers. Int. J. Photoenergy, 3: 187-191.

18. Mendham, J., R.C. Denney, J.D. Barnes and M.J.K. Thomas, 2000. Vogel's Textbook of Quantitative Chemical Analysis. 6th Edn., Prentice Hall, New Jersey, USA.

19. Muruganandham, M. and M. Swaminathan, 2004. Photochemical oxidation of reactive azo dye with $UV-H_2O_2$ process. Dyes Pigments, 62: 269-275.

20. ppenlander, T., 2003. Photochemical Purification of Water and Air. Wiley-VCH, Weinheim, Germany.

21. Petri, B.G., R.J. Watts, A.L. Teel, S.G. Huling and R.A. Brown, 2011. Fundamental of ISCO Using Hydrogen Peroxide. In: In situ Chemical Oxidation for Groundwater Remediation, Siegrist, R.L., M. Crimi and T.J. Simpkin (Eds.). Springer, New York, pp: 35-36.

22. Ren, J., Q.W. Ma, H.H. Huang, X.R. Wang, S.B. Wang and Z.Q. Fan, 2010. Oxidative degradation of microcystin-LR by combination of UV/H_2O_2. Fresenius Environ. Bull., 19: 3037-3044.

23. Riga, A., K. Soutsas, K. Ntampegliotis, V. Karayannis and G. Papapolymerou, 2007. Effect of system parameters and of inorganic salts on the decolorization and degradation of Procion H-exl dyes. Comparison of H_2O_2/UV, Fenton, UV/Fenton, TiO_2/UV and $TiO_2/UV/H_2O_2$ processes. Desalination, 211: 72-86.

24. Seraghni, N., S. Belattar, Y. Mameri, N. Debbache and T. Sehili, 2012. Fe (III)-citrate-complex-induced photooxidation of 3-methylphenol in aqueous solution. Int. J. Photoenergy, 10.1155/2012/630425

25. Tang, W.Z., 2003. Physicochemical Treatment of Hazardous Waste. Lewis Publisher, New York, USA.

26. Xu, B., N.Y. Gao, H. Cheng, S.J. Xia, M. Rui and D.D. Zhao, 2009. Oxidative degradation of dimethyl phthalate (DMP) by UV/H_2O_2 process. J. Hazard. Mater., 162: 954-959.

Chapter 4

Measurements of Methane Emissions from Natural Gas Gathering Facilities and Processing Plants: Measurement Methods

J. R. Roscioli[1], T. I. Yacovitch[1], C. Floerchinger[1], A. L. Mitchell[2], D. S. Tkacik[2], R. Subramanian[2], D. M. Martinez[3], T. L. Vaughn[3], L. Williams[5], D. Zimmerle[4], A. L. Robinson[2], S. C. Herndon[1], and A. J. Marchese[3]

[1]Aerodyne Research Inc., Billerica, MA, USA

[2]Department of Mechanical Engineering, Carnegie Mellon University, Pittsburgh, PA 15213, USA

[3]Department of Mechanical Engineering, Colorado State University, Fort Collins, CO 80523, USA

[4]The Energy Institute, Colorado State University, Fort Collins, CO 80523, USA

[5]Fort Lewis College, Durango, CO 81301, USA

ABSTRACT

Increased natural gas production in recent years has spurred intense interest in methane (CH_4) emissions associated with its production, gathering, processing, transmission and distribution. Gathering and processing facilities (G&P facilities) are unique in that the wide range of gas sources (shale, coal-bed, tight gas, conventional, etc.) results in a wide range of gas compositions, which in turn requires an array of technologies to prepare the gas for pipeline transmission and distribution. We present an overview and detailed description of the measurement method and analysis approach used during a 20-week field campaign studying CH_4 emissions from the natural gas G&P facilities between October 2013 and April 2014. Dual tracer flux measurements and onsite observations were used to address the magnitude and origins of CH_4 emissions from these facilities. The use of a second tracer as an internal standard revealed plume-specific uncertainties in the measured emission rates of 20–47 %, depending upon plume classification. Combining downwind methane, ethane (C_2H_6), car bon monoxide (CO), carbon dioxide (CO_2), and tracer gas measurements with onsite tracer gas release allows for quantification of facility emissions, and in some cases a more detailed picture of source locations.

INTRODUCTION

The natural gas industry has undergone a transformation in recent years, in large part due to technological advancements such as hydraulic fracturing and horizontal drilling. These advances have led to increases in domestic natural gas production (EPA, 2014b), although concomitant with this increase has been a rising concern over methane emissions from the entire natural gas system, from the perspective of both environmental impact and a loss of resources or product. Over the past decade, many studies have aimed at quantifying these emissions using a variety of methods, yielding a wide range of emissions assessments (Pétron et al., 2012; Allen et al., 2013; Kar- ion et al., 2013; Bullock and Nettles, 2014; Subramanian et al., 2014; Zimmerle et al., 2014; Harrison et al., 2011; Zavala-Araiza et al., 2014).

The path of natural gas from well to the consumer can be considered in terms of five possible steps: production; gathering; processing; transmission and storage; and distribution. 5 A recent series of studies have investigated CH_4 emissions from each of these activities (Subramanian et al., 2014; Zimmerle et al., 2014; Allen et al., 2013). Presented here is a discussion of the methods used during one such investigation, where tracer release techniques were used to study emissions from gathering and processing (G&P) facilities (Mitchell et al., 2014; Marchese et al., 2014). This approach 10 is similar to that employed in previous field measurements of distribution, production, transmission and storage facilities (Allen et al., 2013; Subramanian et al., 2014; Lamb et al., 2014). Of particular emphases in this report are the measurement approach to the field campaign and the unique emission profiles associated with gathering and processing, illustrating the wide variety of handling, treating, and processing tools at the 15 disposal of the natural gas industry. The G&P field campaign was executed by Aerodyne Research, Inc. (ARI), Carnegie Mellon University (CMU), and Colorado State University (CSU) from October 2013 through April 2014. Mobile laboratories operated by ARI and CMU sampled emissions from a total of 130 G&P facilities across 20 natural gas basins in 13 states, using tracer release methodology, as discussed below. 20 The measurements were performed with cooperation from industry partners, who provided site access and detailed facility data, such as natural gas throughput, gas type, gas composition, equipment inventories, compressor power, age, and inlet/outlet pressures. E_orts were made by the study participants to ensure that the facilities were sampled as found, and the resulting data was assigned a random number such that it 25 cannot be traced back to a specific facility or partner company.

The inherent chemical profile of natural gas from di_erent sources can significantly affect the technological approach that G&P facilities use to prepare the gas for delivery into the transmission pipeline system. In order to sample from the wide range of equipment employed during gathering and processing, the campaign measured emis- sions from facilities associated with a variety of types of gas, such as gas with low- and high-C_2+ hydrocarbon content (here referred to as dry and wet gas, respectively), as well as sour (high sulfur and/or CO_2 content) and sweet gas sources (low sulfur and/or CO_2 content). More detailed information about site selection is presented by Mitchell et al. in the associated

Measurements paper (Mitchell et al., 2014). These facilities handled natural gas derived from a variety of origins, including shale, coal-bed, and conventional wells. In many cases, the emission profiles associated with these facilities reflect the equipment used to prepare the natural gas (EIA, 2006; Kidnay et al., 2011). For example, the first step during gathering is often passage through gathering lines and a compressor (gathering) station. One of the primary purposes of gathering facilities is to collect and compress the input stream of gas to pipeline pressures, usually _ 800 psi. This requires the use of compressors and associated equipment, for which there are multiple possible emission sources such as compressor seals, natural gas-driven pneumatic devices and engine exhaust. Frequently gathering facilities will also remove water from the gas stream using dehydration trains, which provide more possible emissions points. Following gathering, sweet, dry gas can typically be easily conditioned and sent to the distribution network. However, gas that is sour, wet, or with a high water content requires significant subsequent processing, such as the removal of natural gas liquids using forced extraction, and sometimes a dehydration step to further remove water (Kidnay et al., 2011; Jumonville, 2010). These relatively complex structures can involve distillation columns, turboexpanders, separators, compressors, pneumatic devices and heat exchangers, all of which can emit CH_4 either through minor fugitive components or venting. Finally, extracted natural gas can have high CO_2 and/or H_2S content (i.e. sour, especially in coal-bed methane and some shale gas re25 gions), which requires amine treating (frequently collocated with other gas processing or compression facilities) to make it distribution-ready (Kidnay et al., 2011). Again, this equipment and additional processing adds to the number of possible emission sources. Presented in the second half of this paper are examples of the unique chemical profiles associated with the gathering, treatment and processing systems utilized by the natural gas industry. In the process of measuring CH_4 emission rates, these signatures can provide important information about contributions from specific methane sources on site.

CHALLENGES IN MEASURING EMISSIONS FROM NATURAL GAS FACILITIES

The necessity for emissions measurements at natural gas facilities is two-fold: (i) as an assessment of the impact of facility operation upon regional and national air quality and climate (EPA, 2014a) and (ii) to quantify losses due to normal operation or identify large emission sources. In the case of (i), measured emissions provide an opportunity to compare to national estimates, and assess the overall impact of the natural gas supply chain on CH_4 emissions in the US (Marchese et al., 2014; Subramanian et al., 2014). In the case of (ii), these measurements aid the natural gas industry in minimizing product losses.

Bottom-Up Approaches

Several approaches have been utilized to observe emissions at industrial facilities. In some cases, a bottom-up approach is employed, wherein the magnitudes of emissions from individual components are directly measured and then added together to estimate the facility-level emission rate (FLER) (Subramanian et al., 2014; Harrison et al., 2011). This makes use of stack test data, manufacturer data, emission factors, engineering estimates, activity factors and onsite measurements. These onsite measurements can take many forms, such as acoustic emission detection, which quantifies leaks through suspected leak points such as valves, and Hi-Flow® sampling, which can accurately determine emission rates from a variety of fixtures. While these methods are widely used and are capable of many measurements in a short time, they are not applicable to all possible emission sources due to the number and accessibility of fixtures within facilities (Subramanian et al., 2014). This issue is particularly relevant at large process- ing and treating plants, where the inability to measure emissions from a large number of components could lead to an asymmetric bias in the reported FLER. In addition, in order to accurately scale bottom-up studies to nationwide (or even regional) estimates, care must be taken to ensure that the sampled population, which is typically small, accurately represents the national or regional inventory of facilities.

Optical gas imaging (e.g. infrared cameras such as FLIR®) is a method by which leaks can be identified by using real-time infrared imaging. This method provides a high duty cycle – dozens of fixtures within a facility can be investigated per hour – and large emitters can be readily identified. It is often used in conjunction with the above methods to locate possible leak sources. However, because the method does not measure CH_4 concentrations or flow rates, it does not quantify the emission magnitudes. It nonetheless serves as a powerful qualitative tool in leak detection, and is therefore leveraged in this study to identify suspected emission points at each G&P facility.

Top-Down Approaches

Top-down estimates aim to quantify methane emissions from a particular geographic region. These results can then be compared to inventories constructed from bottomup measurements. Two top-down approaches are commonly used for determining regional methane emissions: mass-balance flights and fixed sensors fields (Zavala- Araiza et al., 2014). The mass-balance flight method, exemplified in several recent oil and gas basin studies (Karion et al., 2013; Pétron et al., 2013, 2012), uses upwind and downwind transects to capture emissions from a bounded region. This area can be as small as an individual facility, or as large as an entire basin. Under favorable meteorological conditions, such measurements can potentially estimate emissions from a large area with a single flight, but these techniques are costly and provide little to no source specificity. This lack of source-specificity makes it especially diffcult for top-down studies to determine the relative emissions from various activities within the industry (i.e. from gathering, processing, transmission, or production), or even differentiate between emissions from different industries, such as natural gas vs. feedlots vs. farming op- erations vs. natural emissions. In addition, due to costs, these studies have a limited number of samples over a short duration (hours), and therefore may not be representative of actual emissions when extrapolated and compared with annual nationwide inventories.

Top-down estimates of regional emissions are also commonly performed using meteorological transport simulations in combination with a network of fixed sensors (McKain, 2012; Bullock and Nettles, 2014; Zavala-Araiza et al., 2014). Such methods can leverage preexisting sensor networks with data available 24 hday1. However, the interpretation of sensor data for emissions measurements is highly dependent upon at mospheric modelling, with large uncertainties (Nehrkorn et al., 2010; Draxler and Hess, 1998, 1997).

Tracer Release Approach

Because the goal of this study was to develop an understanding of the total emissions from individual G&P facilities, and to use these measurements to estimate total national emissions from natural gas gathering and processing (Marchese et al., 2014), the measurement approach described here uses an established measurement technique called tracer flux ratio (or tracer ratio). It has previously been demonstrated that the tracer ratio method can quantify the total emissions from industrial sites (Lamb et al., 1995; Allen et al., 2013) and landfills (Czepiel et al., 1996; Mosher et al., 1999). The strengths of the method are that it does not require theoretical modeling, can measure facility-wide emissions, and under the proper conditions can be useful in identifying large sources within a facility. The tracer ratio method has been shown to e_ectively and accurately yield the total emissions from many small sources within a large area, where measurements of individual leak rates would be challenging (Shorter et al., 1997; Mosher et al., 1999; Subramanian et al., 2014; Lamb et al., 1995). It therefore allows for facility-level emission rates (FLERs) to be determined for large facilities such as processing and treatment plants, where a multitude of possible emissions sources exist that may not be accessible or quantifiable using bottom-up approaches. For this study, the method is applied to quantify total facility-level methane emission rates (fugitive, venting and combustion) at natural gas processing plants, treatment facilities and midstream compressor stations.

Conceptually, the tracer release method is based upon the simple relation that the 5 downwind concentration enhancement of gas X above ambient background, $É[X]$, is directly related to the flow rate at its source, F_X:

$$\Delta[X] = \alpha \cdot F_X$$

(1)

The relation between these two quantities is determined by . The coeffcient α is a complicated function of meteorological information, such as wind speed, wind his tory, turbulence, solar irradiance, temperature, boundary layer height, local topography and downwind distance. In principle this information can be estimated using, for example, a Gaussian dispersion model (Beychok, 2005). Such models have had success in qualitatively reproducing measured plume data, but frequently lack the precision and accuracy required for this study, especially in areas with complex terrain and meteorol ogy.

The tracer release method provides an empirical means to bypass the need for determining   (Lamb et al., 1986, 1995). By deploying a known flow of tracer gas located physically near a CH_4 emission source, the downwind tracer concentration enhancement (above background), $\Delta[T]$, downwind CH_4 concentration enhancement (above 20 background), $\Delta[CH_4]$, and tracer flow rate, FT , become measurable quantities. The ratio of the two downwind concentrations is then equal to the ratio of flow rates:

$$\frac{\Delta[CH_4]}{\Delta[T]} = \frac{\alpha \, F_{CH_4}}{\alpha \, F_T} = \frac{F_{CH_4}}{F_T}$$

(2)

where FCH_4 refers to the flow of CH_4 from the facility. Because concentrations $\Delta[CH_4]$ and $\Delta[T]$ are measured, and FT is known, FCH_4 can be determined without the need for 25 detailed information about .

The underlying assumption in this technique is that the tracer release point is located close enough to the unknown emission source that both gases experience the same dilution factor _. This separation distance becomes less important as the concentration measurement (aboard a mobile platform) moves further downwind. However, when 5 the separation distance is of the same order as the downwind distance, the _ values associated with CH_4 and T are expected to be significantly different. Under ideal circumstances, the tracer is collocated with the emission source, and their concentrations are measured far downwind in stable meteorological conditions. In practice this is not always possible due to facility size, interfering methane sources, road access, or vary ing winds.

To mitigate these issues, this study made use of a *dual* tracer release technique (Allen et al., 2013), where two different tracer gases, in this case N_2O and C_2H_2 are released from di_erent locations within the facility, bracketing the onsite equipment, as shown in Fig. 1. The use of a second tracer has two important advantages over single 15 tracer measurements. First, closer downwind measurements (50–200m downwind) afford a refined assessment of an emission source location based upon the position of its CH_4 plume relative to each tracer plume. Second, when conducting mixed plume characterization in the far-field (downwind), where $N_2O \sim C_2H_2 \sim CH_4$, the second tracer becomes an internal standard to the measurement. The use of two known tracer 20 gas flow rates and an observed downwind molar ratio then provides an empirical measure of the uncertainty for every plume.

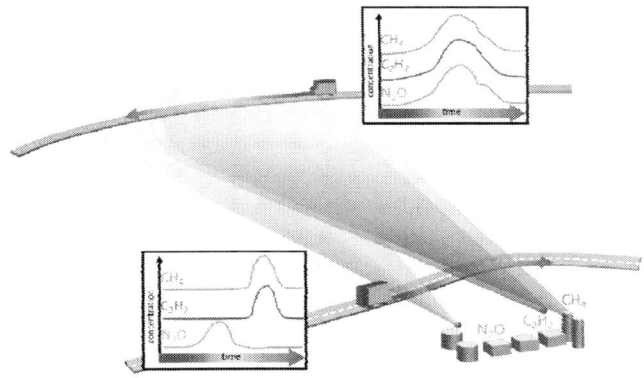

Figure 1: Schematic of dual tracer release technique. At distances far down-wind (top), both tracers and CH_4 are spatiotemporally overlapped. At distances closer to the facility, the spatial position of the CH_4 plume relative to the two tracer plumes can indicate the location of an emission vector onsite with sub-facility resolution.

Understanding and Optimizing Data Quality

In the context of the two possible transect scenarios depicted in Fig. 1 (spatially over25 lapping plumes vs. spatially separated plumes), it is important to qualitatively understand what measurement conditions (tracer separation, transect distance, meteorology) yield these two results. This can be developed using Gaussian dispersion modeling as a guide (Beychok, 2005). As a rule of thumb, for typical mid-day atmospheric conditions (stability classes A, B, or C) and downwind distances (100–3000 m), the horizontal width of a plume that is propagating according to Gaussian dispersion is ~ 20–50% of the distance that it has traveled from it source. That is, the ratio of plume width to downwind distance is 0.2–0.5, where low wind conditions yield wider plumes 5 (~ 0.5), and high wind conditions yield narrower plumes (~ 0.2). A plume observed 1000m downwind of its origin, for example, is typically 200–500m wide.

If the plume widths of two gases being measured downwind (e.g. CH_4 and N_2O) are much larger than the separation of their sources, the plumes will generally be codispersed, or spatially overlapping. Therefore the ratio of the distance between emis10 sion sources to the downwind transect distance must be less than 0.2–0.5 in order to achieve co-dispersion. If, for example, the separation between an N_2O tracer and a CH_4 source is 100 m, the downwind distance required to observe the onset of co-dispersion is > 500m in high winds, and > 200m in low winds. Alternatively, if local road access limits the downwind distance to 200–500 m, the N_2O tracer must be placed within 100m of the suspected CH_4 emission source.

This same rule-of-thumb approach can be applied to cases where a nearby CH_4 source, such as a wellhead, may interfere with the FLER measurement at a G&P facility. In these cases, the downwind transect must be *close* enough that the interfering plume width is smaller than its separation from the G&P facility. For example, if the distance 20 between a wellhead and facility is 50m, downwind transects must be less than 100– 250m in order isolate end exclude the wellhead plume from the FLER estimate.

When the second tracer is used as an internal standard, it can serve to quantify the uncertainty of the measurement. This uncertainty decreases when the two tracer plumes are spatially overlapping, as compared to cases where the plumes are separated. Because this precision reflects the uncertainty in the FLER, efforts are made by the study team to maximize the codispersion of methane and tracer plumes. In light of the above discussion, this can be achieved by attempting to place one or both tracers near the dominant suspected emission source at a facility, if one exists. When these conditions are met, the down-wind distance required to observe co-dispersion is reduced, thereby increasing the instrumental signal-to-noise and further separating any possible interfering sources.

Initial placement of the tracers at opposite ends of the facility allows for early transects to identify suspected methane emission locations. In some cases, the observed methane 5 plume will appear covariant with one of the two tracers, indicating that the dominant methane emitter is in the vicinity of that tracer. In many cases, however the methane plume is observed between the two tracer plumes. In this scenario, one (or both) of the tracers is typically moved such that its plume is spatially overlapping the methane plume. This process is iterated multiple times over the course of the measure ment in order to yield plumes that exhibit high degrees of CH_4-to-tracer correlation.

While two tracers act as an internal standard in the horizontal plane, a complicating factor unique to some large facilities (e.g. processing plants and larger gathering facilities) studied here is the presence of flares and/or engine exhaust stacks, some of which can be over 20m tall. A simple rule-of-thumb approach as used above is hampered here by both buoyant plume rise effects and plume reflection off of the ground. These calculations indicate that in strong wind conditions (i.e. high atmospheric stability classes, such as in winds above $5ms^{-1}$), the measured emission rate determined from close transects can be biased considerably low, depending upon the fraction of the emission coming from elevated positions. In wind conditions below $5ms^{-1}$, the dispersion is large enough that the bias is lessened to 0–50 %. To minimize this bias, plumes were obtained as far downwind as possible, and at several processing plants a tracer was emitted at an elevated position such as the side of a demethanizer column or stack. The impact of the bias upon the overall data set and resulting conclusions is discussed in more detail in the accompanying Measurements paper (Mitchell et al., 2014).

Auxiliary Species

The study team also used measurements of other species, CO, CO_2 and C_2H_6, to aid in identifying and attributing methane emissions to targeted G&P facilities. For example, engine exhaust from reciprocating engines and turbines that power compressors at many 5 natural gas facilities will contain CO and CO_2. This enables potential differentiation between emissions of G&P equipment and those emanating from nearby well pads (which typically do not include combustion sources, or emit much smaller amounts of CO and CO_2). Similarly, amine treatment systems serve as non-combustion sources of CO_2 and are easily distinguishable from other facilities (Rochelle, 2009; EIA, 2006; 10 Kidnay et al., 2011).

Ethane measurements serve multiple purposes within the context of this study. First, the presence of ethane associated with methane in downwind plumes indicates that some fraction of the methane is of thermogenic, rather than biogenic, origin. The ability to distinguish between these sources is especially important in farming and ranch15 ing regions, where livestock emissions can be a substantial source of CH_4. Second, the observed ethane-to-methane ratio (E =M ratio) in a downwind plume can serve as a unique identifier of a facility of interest. It can therefore be used to differentiate a particular emission source from others in the area. Finally, variations in ethane content over close transects can indicate active distillation or other processing present 20 onsite. The utility of these measurements will be explicitly illustrated via examples in the Sect. 6.

LABORATORY AND INSTRUMENT DETAILS

The two mobile laboratories used in this study were operated by Aerodyne Research, Inc. (Herndon et al., 2005) and Carnegie Mellon University (Subramanian et al., 2014). 25 Both mobile laboratories contain of a variety of spectroscopy-based gas-detection instruments, which sample the ambient air from an inlet mounted on the front of the vehicle. In the case of the Aerodyne Mobile Laboratory, 3 ARI direct-absorption quantum cascade laser (QCL) spectrometers (Jiménez et al., 2005; Yacovitch et al., 2014; McManus et al., 2005) operating at 20–40 Torr are employed in series to detect CH_4, C_2H_6, CO, N_2O, and C_2H_2. To detect CO_2, a non-dispersive IR (NDIR) LiCOR® instrument is used. 5 In this work, the QCL spectrometers were operated in series, with flow rates through the instruments of _ 10 SLPM. This flow rate afforded a time response that is < 1 s. The NDIR instrument drew a small flow from the inlet line before the air sample entered the QCLs. The QCL spectrometers report mixing ratios of all species in parts per billion by volume (ppbv), while the NDIR instrument reports CO_2 in parts 10 per million by volume (ppmv). On the Carnegie Mellon Mobile Laboratory, CH_4 and C_2H_2 are measured using a Picarro Cavity

Ringdown Spectrometer (Crosson, 2008; Rella et al., 2009) running at 4–5 Hz, while C_2H_6, N_2O, and CO were measured using an ARI Dual QCL spectrometer operating at 1 Hz. Detection limits of all instruments are listed in Table 1. Except for practically limiting the minimum detectable concentration of certain species, the differences in equipment manufacturer and sensitivity do not affect the results of the measurements. In addition to the concentration information, both mobile laboratories record their location, bearing, and heading using Global Positioning Systems (GPS, Garmin® 76 and Hemisphere GPS Compass® for the ARI laboratory, Airmar® for the CMU laboratory). A small meteorological station (Airmar® 20 200WX or LB150) is also mounted on a boom at the front of the vehicle to record true wind speed, true wind direction, and GPS location. Along with the mixing ratios, this information is recorded at 1 s intervals on a main onboard acquisition computer, where all of the acquired data are visualized in real time and can be overlaid on maps.

Table 1. Instruments and sensitivities for measured species on Aerodyne and CMU Mobile Laboratories

Instrument	Species Detected	Sensitivity
Aerodyne Mobile Laboratory		
Aerodyne Dual QCL	CH_4	1 ppb
	C_2H_2	200 ppt
Aerodyne Mini QCL	C_2H_6	100 ppt
Aerodyne Mini QCL	N_2O	100 ppt
	CO	100 ppt
Li-Cor NDIR	CO_2	500 ppb
Carnegie Mellon Mobile Laboratory		
Picarro CRDS	CH_4	3 ppb
	C_2H_2	600 ppt
Aerodyne Dual QCL	C_2H_6	100 ppt
	N_2O	100 ppt
	CO	100 ppt

Both laboratories are accompanied by a tracer release vehicle (i.e. pickup truck) to 25 facilitate the storage, set up and release of the N_2O and C_2H_2 tracers. Tracer gas bottles are stored on the bed of the truck, along with flow control systems and associated valves, tubing and telemetry systems. Polyethylene tubing for each tracer is rolled out from the pickup truck up to 200m to the intended release location, where the end of the tube is attached to onsite equipment or is placed on a tripod. For both laboratories, tracer flow rates are controlled by Alicat® MC-series mass flow controllers. The mass flow rates are recorded via RS232 to an onboard computer in the vehicle.

In addition to the tracer gas flow systems, three portable meteorological stations (Airmar® 200WX) are deployed on tripods, sometimes serving as physical supports for 5 the tracer release tubing. They are capable of recording GPS, true wind direction and wind speed with 1 s resolution. Each unit broadcasts that information wirelessly or via an RS232 cable at 1 Hz to a computer onboard the tracer release vehicle, where it is recorded and displayed for observation by the tracer release personnel to advise the mobile laboratory as needed. When considered in the context of tracer placement, the 10 wind data can immediately inform mobile laboratory personnel whether a tracer is being deployed in an area onsite that is not well-ventilated. If this is the case (frequently due to the local wind currents near buildings) the tracer can then be moved to allow it to be carried downwind by the larger regional wind mass. This information also provides a crude wind field for later analysis to better understand the sources of error and 15 uncertainty in tracer release methods.

Calibrations and Ranges

In both laboratories, the inlet was periodically overblown (injected with a flow larger than the intake flow) with ultrazero air (AirGas® or Praxair®) to zero the instruments, typically every 15 min for 30 s. Because CH_4 and N_2O are present in background ambient air 20 (1900 and 325 ppbv, respectively), zeroing events also serve as an approximate check of those instrument calibrations. Full instrument calibrations were performed several (4–5) times over the course of the measurement campaign using calibration standards. For these dilution calibrations, a controlled mass flow of calibration gas is released into a known zero-air flow, and the resulting mixture is overblown into the inlet. The mixture 25 is changed by varying the calibration gas flow using either a series of critical orifices or mass flow controllers (Alicat® MC Series). The results of these calibrations changed less than 5% over the course of the campaign. The mass flow controllers onboard the tracer release vehicle are also periodically calibrated using a NIST-traceable Dry-Cal® flow meter.

FIELD IMPLEMENTATION

In practice, when the mobile laboratory arrived at a facility, a safety meeting was con- 5 ducted with the facility supervisor, after which the tracer release apparatuses were set up. The tracer positions were decided upon after discussion with the supervisor regarding likely emission sources (near compressors, dehydrators, tanks, etc.), a cursory survey with infrared imaging, consideration of the current wind conditions, site size and safety issues, and sometimes after performing an initial drive within facility 10 boundaries. After setup, the tracer gases were released and the mobile laboratory was deployed downwind. Constant communication was maintained either over CB radio or cellular phones. During this period, an additional study team member ("the onsite observer") surveyed the facility with an infrared camera, inventoried facility components, and recorded relevant information such as facility throughput, equipment counts and 15 motor, engine, or turbine horsepower. In many cases the identification of emission sources by survey of the facility using infrared imaging agreed with or informed the results of close-pass plume transects. If the mobile laboratory detected CH_4 plumes that were spatially separated from the tracer plumes, one or both tracers were moved to maximize co-dispersion with CH_4. When possible, onsite ethane-to-methane ratios 20 were measured by driving the mobile laboratory within fenceline immediately downwind (< 25 m) of onsite equipment, for future comparison with partner company gas chromatograph (GC) data.

After acquiring enough downwind plumes (a target of 10) to provide a statistically meaningful time-averaged FLER and uncertainty, the mobile laboratory returned on 25 site, and the tracer release hardware was packed. Usually at least two facilities were surveyed daily, and sometimes as many as four, depending upon wind conditions, time, and the locations of nearby facilities. Because of their size and scale, a full day was reserved to sample emissions from processing facilities.

PLUME TYPES AND ANALYSIS METHODS

There are multiple ways in which downwind tracer plumes can be analyzed, depending 5 upon the plume intensity and spatial overlap between the tracer and CH_4 plumes (Subramanian et al., 2014). Figures 2–5 show the four possible plume types observed during the G&P campaign.

Dual Correlation

The ideal scenario occurs when the measurement transect is far enough downwind 10 of the facility that the CH_4, N_2O, and C_2H_2 plumes are spatially overlapping. The resulting measurements of concentration vs. time exhibit a high degree of covariance between species, as shown in the top panel of Fig. 2. Analysis of these "dual correlation" plumes consists of plotting the concentration of one species vs. another, and performing a linear orthogonal distance regression (ODR) fit as shown in the bottom 15 panels of Fig. 2. This regression analysis is performed for CH_4 vs. N_2O, CH_4 vs. C_2H_2, N_2O vs. C_2H_2, and C_2H_6 vs. CH_4. From these linear regressions, the slope indicates the ratio of concentrations of the two gas species (for use in Eq. 2), and R^2 indicates the degree of correlation. These values are recorded for use in determining whether the plume meets the acceptance criteria for the CH_4 emission rate to be considered 20 valid. If the R^2 values derived from fits of CH_4 vs. N_2O, CH_4 vs. C_2H_2, and N_2O vs. C_2H_2 are all greater than 0.75, and the tracer ratio ($[C_2H_2] = [N_2O]$) is within a factor of 1.5 of the known tracer flow rate, the plume is a candidate for dual correlation analysis. The choice of acceptable R^2 and tracer ratio were based upon values at which further relaxation of the criteria would alter the uncertainty and accuracy of the FLER measurement (Mitchell et al., 2014).

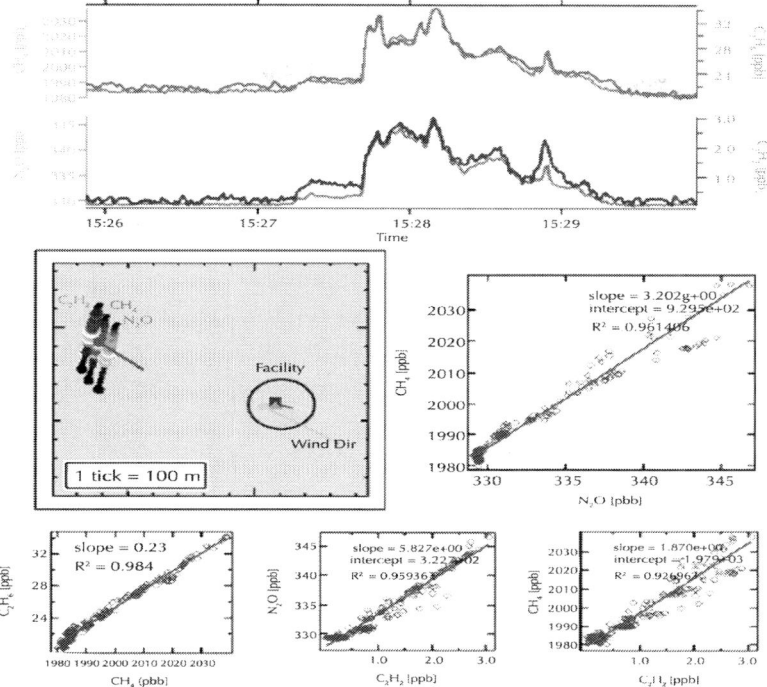

Figure 2: Example dual-correlation plume from a natural gas facility. Top panel: time trace of CH_4, C_2H_6, N_2O, and C_2H_2 concentrations, showing high temporal correlation. Center left panel: map of tracer location (right side) and transect location (left side) during the course of the plume. Red, blue, and green weighted lines correspond to CH_4, C_2H_2, and N_2O intensities, respectively, during the transect, spatially o_set for clarity. Thin lines point into the wind at the mobile laboratory (red) and at the facility (light blue, pink, and yellow). Blue square and green triangle indicate C_2H_2 and N_2O release locations, respectively. Lower panels: correlation analysis of C_2H_6 vs. CH_4, N_2O vs. C_2H_2, CH_4 vs. C_2H_2 and CH_4 vs. N_2O.

Dual area

In certain circumstances, wind conditions along with local road access and intervening CH_4 5 sources prevent the ability to get far enough downwind for the tracer gas and CH_4 plumes to become spatially overlapped. However, transects may still be performed closer to the facility (~ 50 ~ 500 m) such that all three species will be observed. As illustrated in the example shown in Fig. 3, under these circumstances correlation diagrams do not provide useful information about the ratio of species (bottom panels). In these 10 cases a "dual area" technique is used, where the analysis must rely on the integrated area of each species' plume over the time of the transect. Here, the deviation of the species' mixing ratios from ambient conditions must be considered, rather than the raw integrated intensity. This point is particularly relevant for CH_4 and N_2O, whose ambient concentrations are ~ 1900 and ~ 325 ppb, respectively. In the analysis of the data, the 15 baseline (non-plume) mixing ratio was determined by fitting a line through the average of several data points immediately before the plume transect began and the average immediately after the transect ended. The fit line was then subtracted from the data to yield a baseline-corrected plume. This accounted not only for background concentrations (e.g. 1900 or 325 ppb), but also any minor baseline drift that may have occurred 20 over the course of the transect. The quality of the baseline fit was visually confirmed and corrected if it did not accurately represent the true baseline. For the plume to be considered a candidate for dual area analysis, the ratio of areas of the C_2H_2 and N_2O plumes must be within a factor of 2 of the known tracer flow rates.

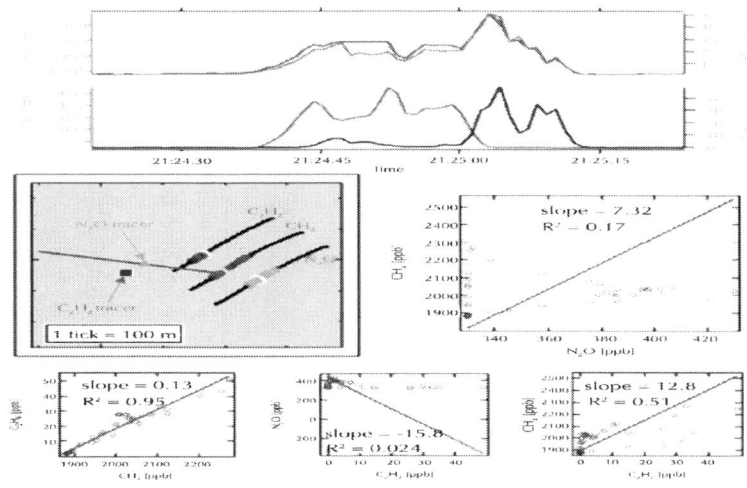

Figure 3: Similar to Fig. 2., illustrating dual area-type plumes. Top panel: time trace of CH_4, C_2H_6, N_2O, and C_2H_2 concentrations, showing high temporal

correlation. Center left panel: map of tracer location (right side) and transect location (left side) during the course of the plume. Red, blue, and green weighted lines correspond to CH4, C_2H_2, and N_2O intensities, respectively, during the transect, spatially o_set for clarity. Thin lines point into the wind at the mobile laboratory (red) and at the facility (light blue, pink, and yellow). Blue square and green triangle indicate C2H2 and N2O release locations, respectively. Lower panels: correlation analysis of C2H6 vs. CH4, N2O vs. C2H2, CH4 vs. C2H2 and CH4 vs. N2O. Note the lack of correlation in lower left and center panels, indicating that the analysis must rely on an area method.

Single Correlation

25 In scenarios where the CH_4 mixing ratio was highly correlated with only one of the two tracers, a "single correlation" analysis was performed, as shown in Fig. 4. This approach corresponds to that originally used by Lamb et al. in early demonstrations of the tracer release method (Lamb et al., 1995). The need to use the single correlation technique can be the consequence of several possible measurement conditions: (i) one of the tracers is placed geographically close to the dominant emitter within the facility 5 (e.g. a compressor or large fugitive source), (ii) the site is emitting a tracer species (i.e. C_2H_2 during certain combustion processes), forcing the measurement to become single-tracer only, or (iii) the plume transect is far enough downwind (frequently > 2 km) that one of the tracer species' mixing ratio is at or below the instrumental detection limit. In single correlation cases, correlation analysis is performed for both tracers, but only 10 the well-correlated tracer serves to provide the true CH_4 emission rate. For a plume to be a candidate for single-correlation analysis, the R^2 value derived from the linear regression fit of CH_4 to one of the two tracers must be greater than 0.75.

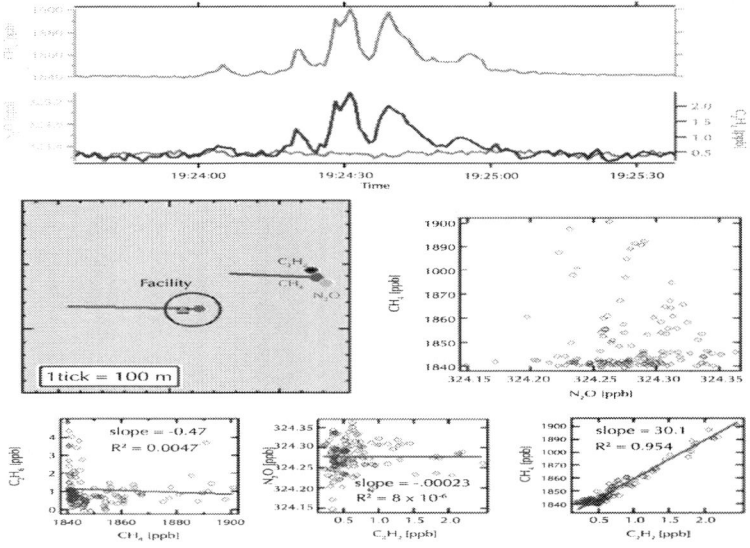

Figure 4: Example of a single-correlation plume (CH_4 correlation with C_2H_2). Top panel: time trace of CH_4, C_2H_6, N_2O, and C_2H_2 concentrations, showing high temporal correlation. Center left panel: map of tracer location (right side) and transect location (left side) during the course of the plume. Red, blue, and green weighted lines correspond to CH_4, C_2H_2, and N_2O intensities, respectively, during the transect, spatially o_set for clarity. Thin lines point into the wind at the mobile laboratory (red) and at the facility (light blue, pink, and yellow). Blue square and green triangle indicate C_2H_2 and N_2O release locations, respectively. Lower panels: correlation analysis of C_2H_6 vs. CH_4, N_2O vs. C_2H_2, CH_4 vs. C_2H_2 and CH_4 vs. N_2O.

Linear Combination of Tracer Plumes

In certain circumstances, unique tracer placement, road access and wind conditions 15 allow for intermediate-distance transects where the CH_4 plume profile is not well correlated with either individual tracer, but is well correlated with a linear combination of the tracer plumes, i.e.

$$\Delta[CH_4] = a \cdot \Delta[N_2O] + b \cdot \Delta[C_2H_2]$$

$$(3)$$

where *a* and *b* are multiplicative coeffcients of the N_2O and C_2H_2 plumes, respectively. 20 Such an example is shown in Fig. 5. This scenario is equivalent to performing two independent single-tracer measurement, where the plumes are overlapping in time. In these cases facility emission rates can be determined by performing a correlation analysis of CH_4 vs. (a $\Delta[N_2O]$+b $\Delta[C_2H_2]$) while adjusting the values of *a* and *b* in Eq. (3). The *a* and *b* values that provide the largest possible R^2 value in the fit are used 25 to determine the CH_4 emission rate associated with each tracer.While the sum of these values serves as the facility level emission rate (FLER), the individual emission rates contain information at sub-facility-level resolution, such as leak or vent magnitudes associated with condensate tanks, compressors or dehydrators.

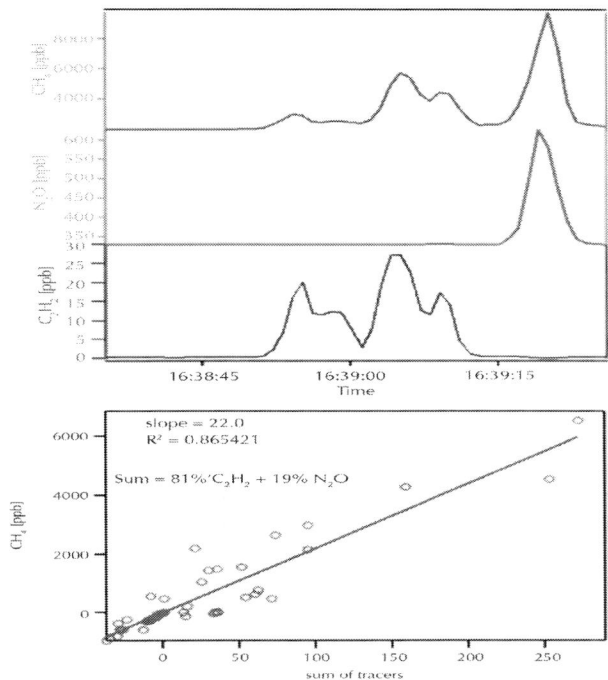

Figure 5: Example of analysis using a linear combination of tracer plumes. Note that N_2O and C_2H_2 are associated with different sections of the CH_4 plume (top). Adding the two tracer plumes in an 81 %/19% combination yields a correlation diagram (below) with high R^2 value (0.87).

This analysis method has also been applied in cases where equipment not associated with the G&P (e.g. a natural gas production well) is present within a facility boundary. 5 In such a case, one tracer is placed at or near the non-associated equipment while the other is placed near a suspected emitter that is part of G&P facility. If the plume from the former tracer is well correlated with the non-associated equipment emission and the plume from the latter tracer is well correlated with the rest of the CH_4 from the facility of interest, then the facility level emission rate can be estimated, even if the CH_4 10 from the non-associated equipment is overlapping with the facility plume.

Implementation of Plume Analysis

Table 2 summarizes the preference of the four analysis methods, their acceptance criteria, the number of accepted plumes that were analyzed using each method, and the measurement variance associated with each plume type.

Table 2: Plume analysis types, preference, criteria, prevalence, and variance

Analysis Type	Preference	Criteria	# of plumes	Variance ($\sqrt{}$Variance)
Dual Correlation	1	— $R^2 > 0.75$: N_2O vs. C_2H_2, N_2O vs. CH_4, C_2H_2 vs. CH4, C_2H_6 vs. CH_4 — Tracer ratio error < 1.5 - E/ M ratio error < 1.5	250	0.04 (0.2)
Dual Area	2/3	— $R^2 > 0.75$: C_2H_6 vs. CH_4 — Tracer ratio error < 2 - E/ M ratio error < 1.5	441	0.14 (0.37)

Single Correlation	3/2	— $R^2 > 0.75$: C_2H_6 vs. CH_4, Tracer vs. CH_4 - E/ M ratio error < 1.5	728	0.09/0.22 (0.3/0.47)
Linear Combination	4	— $R^2 > 0.75$: C_2H_6 vs. CH_4	16	

The large number of plumes observed during the measurement campaign allows for extensive statistical analysis of dual correlation, dual area, and single correlation plumes. This is likely due to the fact that these plumes correspond to a limit where full co-dispersion of the tracers has been achieved, i.e. both tracer plumes are experiencing the same local turbulence by 25 the time they are measured by the mobile laboratory. In addition, no baseline subtraction is required in the dual correlation method, which can be a source of uncertainty depending upon the signal-to-noise exhibited by the plume. The larger variance of the dual area method is likely derived from the lack of co-dispersion of the tracers. In these scenarios, one tracer concentration may be enhanced relative to the other due to the fact that each tracer plume is experiencing di_erent local turbulence en route to the mobile laboratory.

In the case of single correlation plumes, the observed variance is found to be rel- 5 atively small when the downwind tracer ratio (determined using integrated areas) is within a factor of 1.5 of the tracer flow rates (variance of 0.09 in Table 2). Because this variance is less than that for dual area (0.09 vs. 0.14), single correlation analysis is preferred over dual area analysis for these plumes. Notably, the variance increases significantly from 0.09 to 0.22 when including all single-correlation plumes (i.e. with no 10 tracer ratio filter). When the tracer ratio is more than a factor of 1.5 different than the tracer flow rates, the dual area method is then preferred over single correlation analysis. This indicates that although the both tracers are not being used to determine the FLER associated with that plume, filtering by their ratio can still yield more precise results.

Ethane-to-Methane Ratio

Finally, the ratio of ethane to methane in the measured downwind plume can also serve as an acceptance criterion, regardless of plume classification. The amount of ethane in a natural gas mixture can vary from well to well and from one gathering facility to 20 another (Kidnay et al., 2011). As such, the ethane content represents a unique "fingerprint" of a facility, providing a means to identify whether the CH_4 measured in a plume is coming from the facility of interest. In this study, the ethane-to-methane ratio (E =M ratio) associated with a given facility was determined in one of two ways: from partner company GC analysis of the inlet/outlet gas, or from C_2H_6 vs. CH_4 correlation analysis 25 of plumes when the mobile laboratory was onsite (and thus only observing emissions from the facility). While GC analysis data is preferred since it provides a completely independent (and external) check of the methodology, it was not always available on the date of the measurment. When possible, observed E=M ratios of plumes obtained when the mobile laboratory was onsite were compared to the GC data to confirm (or disprove) that the emission composition was in agreement with the GC data.

Both mobile laboratories measured ethane and methane at a 1 Hz sampling rate or faster, allowing for an accurate determination of the E =M ratio of individual plumes. The E5 =M ratio for every downwind plume obtained in the campaign (determined using correlation analysis) was measured and compared to the known ratio from GC analysis (or measured onsite E=M ratio in cases where the GC data was unreliable). A detailed comparison between the observed E=M ratio and that from the inlet GC analysis is presented in the results section. A plume was only accepted for further analysis if 10 the observed ratio was within a factor of 1.5 of the known value. This criterion was suspended in cases where the facility itself was actively changing the ethane content (e.g. from a demethanizer), where the E=M ratio was varying across the facility, or when the downwind C_2H_6 mixing ratio was below the detection sensitivity limit.

Finally, under certain scenarios, a small number of plumes that would be rejected as 15 described above are manually accepted during analysis. These exceptions are possible for one of several reasons. One is that the plume transect is far enough downwind that the tracer or CH_4 plume concentrations are near the detection limit of the onboard instruments. Under such a scenario the correlation analysis may reveal $R^2 < 0.75$, despite the plume being legitimate. Another possible reason for manually accepting 20 a plume is when the E=M ratio is variable across the facility, frequently due to the presence of a high emission point source such as a venting condensate tank. Because condensate tank emissions may exhibit an E=M ratio larger than that of the remainder of the facility, the observed downwind ratio may be variable, even on the timescale of a single plume.

RESULTS

In this section, we present results from a number of case studies that illustrate the capabilities of the dual tracer release method.

Gathering Facilities

A gathering station serves as a point where multiple natural gas sources (wells) are combined to produce a high pressure stream of gas. These facilities typically include equipment such as inlet separators to remove liquid phase water and condensate 5 (C5+), if present, and systems for pipeline maintenance activities (e.g. "pigging"). Compression at these facilities is accomplished by a series of 1 to 20 individual compressors powered by electric motors, reciprocating engines or gas turbines with total engine powers ranging from 500 to 25 000 HP, depending on the inlet gas pressure and total gas throughput (Mitchell et al., 2014). Gathering stations also typically contain conden10 sate storage tanks, produced water storage tanks, and other gas handling equipment including pneumatic valves (often powered by natural gas) and gas metering systems. If the gas has a high water content, glycol dehydration systems are also frequently present to dry the gas (Goetz et al., 2014; Kidnay et al., 2011).

There are three main sources of continuous emissions from these facilities. First, 15 compressors can serve as significant sources of CH_4 via both fugitive leaks as well as through seals in the compressor housing. In the case of wet compressor seals, it should be noted that the primary emission route is due to absorption of methane into the seal fluid at high pressure, followed by exposure of the fluid to ambient pressure, where the methane is routed through a vent to atmosphere (EPA, 2006). Second, 20 because the natural gas is typically under high pressure, fugitive and vented emissions may occur at the facility, including from continuous-bleed natural gas pneumatic devices, dehydration units, and a variety of flanges and valves. Third, methane slip (i.e. unburned methane in engine exhaust gases) through onsite combustion sources such as engines and turbines can be a source of CH_4, depending upon a wide variety 25 of combustion characteristics. The relative importance of this emission source to the FLER is discussed in the associated Measurements report (Mitchell et al., 2014) and in previous studies of combustion emissions in natural gas transmission and storage (Subramanian et al., 2014). Similarly, methane and other unburned hydrocarbons are present in flare emissions, and may vary greatly depending upon the flare combustion effciency (Torres et al., 2012).

Some intermittent methane emission sources may also be found at gathering facilities, such as intermittent-bleed natural gas-driven pneumatic controllers, produced 5 water tanks, and condensate tanks. Of particular importance to the associated Measurements paper (Mitchell et al., 2014), produced water and condensate tanks may *transiently* emit CH_4, C_2H_6 and higher hydrocarbons from thief hatches or other pressure relief valves attached to the tank. Because of the nature of the liquids stored in them, i.e. long-chain hydrocarbons, the ethane to methane ratio observed from a con10 densate tank can be much higher than the natural gas composition entering or exiting the facility. However, these units may sometimes also serve as venting release points for equipment onsite, in which case the E =M ratio will be very similar to that of the inlet stream.

An example of an emission rate measurement from a compressor station (C station) is shown in Fig. 6a. Similar to the example plume shown in Fig. 2, this plume as accepted as dual-correlation (R^2 = 0.998, tracer ratio error = 1.05, E=M ratio error = 1.4). In this case, the methane and ethane signals are strongly correlated with both tracers at a distance of 1600m downwind of the facility. Note that inclusion of the CO and CO_2 in the analysis indicates that both of these gases are also being emitted from 20 the facility, likely due to combustion. While this plume alone can provide an accurate determination of the FLER from the facility, even more information can be extracted by also investigating transects from only 100m away, shown in Fig. 6b (a dual-area plume, with tracer ratio error = 0.7, E =M ratio error = 1.5). While such a close transect may not provide as precise of a FLER, we see from the figure that the CO and CO_2 25 signatures are coincident with only a fraction of the methane being emitted, and are not well-correlated with it. This indicates that some, but not all, CH_4 emitted at the facility may be associated with combustion. In this case, the remaining CH_4 emission is likely from other non-combustion sources onsite. At some facilities, such as that shown in Fig. 6c, CO and CO_2 are correlated with a distinct part of the CH_4 plume, indicating the presence of a combustion source that is emitting CH_4 or co-located with one that is, and clearly associated with one section of the facility. Because the goals of the G&P study are to understand both overall emissions and their origins, this type of analysis can aid in understanding the relative role of combustion sources and methane slip in G&P CH_4 5 emissions. In the case of the compressor station associated with the plume in Fig. 6c, the area of the facility with CO, CO_2, and CH_4 emissions is the compressor/engine section, while the area with no $CO=CO_2$ corresponds to other non-combustion sources onsite. Thus, Fig. 6 illustrates the important role that the auxiliary gas measurements (in this case CO and/or CO_2) can play in identifying sources of emissions.

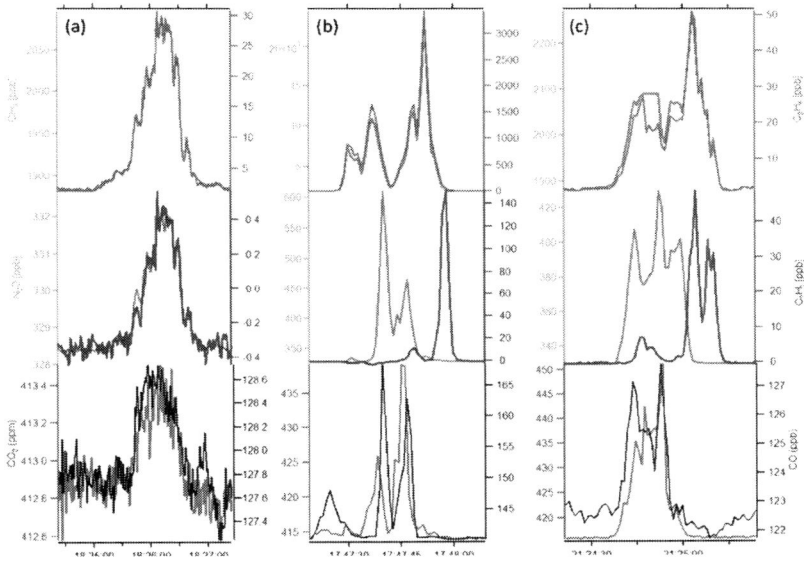

Figure 6: Three exemplary plumes from a gathering station. (a) Far-field plume (1.6 km) showing strong correlation between CH4, C_2H6, N_2O, C_2H_2, CO_2, and CO. (b) Close plume transect (100m away) of same facility, showing loss of correlation and isolation of CO_2 and CO combustion products to a section of the facility. (c) Example of a close plume transect (200m away) showing CO and CO_2 correlation with a component of the CH_4 trace.

Because they are ubiquitous at both production and gathering facilities, it is of interest to this study to understand, and quantify when possible, what fraction of emitted methane is coming from condensate and produced water tanks. Shown in Fig. 7 is an example of the emission profile observed at a compressor facility containing a condensate tank, illustrating another example of the utility of close (< 200 m) transects. In 15 this case, one tracer (N_2O) was placed next to the compressors, while another (C_2H_2) was placed near a battery of 3 condensate tanks. As shown in the transect trace, both of these sources (compressors and tanks) are correlated with their respective tracers, but have very different E =M ratios. Here the relative intensities of the CH_4 plumes associated with the different E=M ratios indicate comparable emission rates between 20 the two sources. As discussed in the associated Measurements paper (Mitchell et al., 2014), the sub-facility spatial resolution afforded by tracer release, along with the measurement of auxiliary species such as ethane, provide the ability to address the contributions of particular equipment, especially condensate tanks, to emissions from G&P facilities. Here, for example, analysis using a linear combination of tracers as described 25 above reveals that the CH_4 emission from the condensate tank represents 50% of the overall CH_4 emission rate from the facility.

While not always the case, it is common to find a larger ethane content in emissions from condensate tanks relative to the inlet gas composition, due to the larger fraction of ethane in the condensate itself. It should be noted that daily temperature variations (producing "breathing" emissions) may change the relative vapor pressures of ethane and methane in the condensate tank, and the filling/emptying schedule of the condensate tank (producing "working" emissions) may alter condensate composition. Both of these activities can therefore change the E =M ratio of the tank emissions over the 5 course of the day.

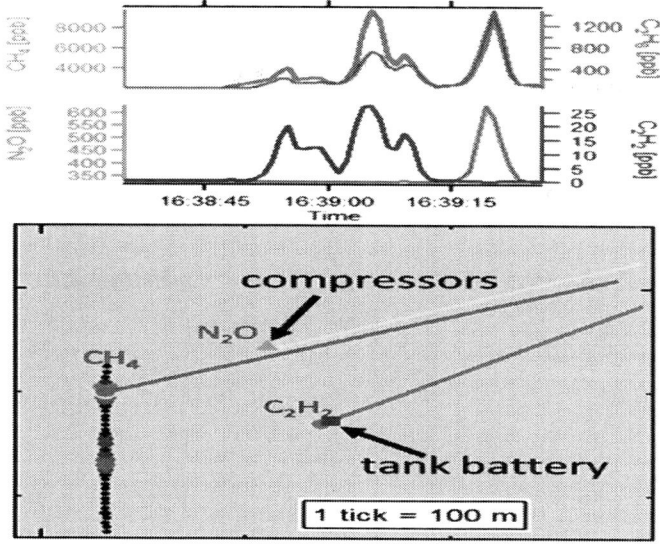

Figure 7: Example of varying E=M ratio during a close transect due to the presence of a condensate tank battery onsite. Note the ~ 2× decrease in the E=M ratio toward the end of the plume.

Amine Treatment

The composition of natural gas often depends upon its geologic origin (or play). To illustrate this effect, we compare emissions from facilities associated with different gas sources: shale and coal bed methane (Whiticar, 1994; Kidnay et al., 2011). Shale gas, 10 tight gas and conventional gas contain varying amounts of ethane and higher hydrocarbons, typically with low levels of CO_2. Coal bed methane, on the other hand, typically contains little ethane and up to 40% CO_2 (Kidnay et al., 2011). This carbon dioxide is particularly interesting since in this case it is not an indicator of combustion. Other combustion sources within the facility can be distinguished by the presence of CO.

If CO_2 is present in high amounts (> 3%), it must be removed from the natural gas prior to transmission and storage. It can be removed from a gas stream by passing the natural gas through a vapor of monoethanolamine or other related amine compounds. This process is called "amine treatment" or "amine scrubbing" (Kidnay et al., 2011; Rochelle, 2009; Bottoms, 1930). The amine binds to the CO_2, and is then regenerated 20 through heating. CO_2 is thus evolved from this process, so the facility's CO_2 emissions relative to CH_4 will be higher than would be expected for a direct leak of the untreated gas. Heating is applied through combustion of excess fuel (natural gas or other easily available source) so CO_2 may sometimes be present along with small amounts of combustion products such as CO and NOx. Amine treatment is also used for the removal of 25 hydrogen sulfide (H_2S), with the main difference being that the H_2S is highly toxic and must be captured or combusted.

Figure 8 contrasts emissions from facilities associated with coal bed methane and shale gas. The facility in Fig. 8a is a coal bed methane treatment plant, without com- pression. The compressor/dehydration facility shown in Fig. 8b has four compressors and is in a shale region with characteristically high ethane content in the gas. The ethane content of the coal bed methane is observed at a molar ratio C_2H_6 =CH_4 = 0.0215 (Fig. 8a), while the shale gas facility emissions have a much higher measured ratio, C_2H_6 =5 CH_4 = 0.164 (Fig. 8b). The CO_2 emissions vary even more greatly between the facilities, at CO_2 =CH_4 = 165 vs. CO_2 =CH_4 = 3.3. The molar ratio of CO_2 to CH_4 in the former facility's emissions (CO_2 =CH_4 = 165) is 4 orders of magnitude higher than the operator-data for the inlet gas (CO_2 =CH_4 = 0.106). For Fig. 8a, at the distances sampled, no other significant combustion products (such as CO) were ob10 served, indicating that the primary source of CO_2 is from the amine treatment process. This information, along with the observed high degree of correlation between CO_2 and CH_4 at intermediate distances (~ 500 m), suggests that the primary CH_4 emission source is located within or near the amine treatment area of the facility.

(a)

(b)

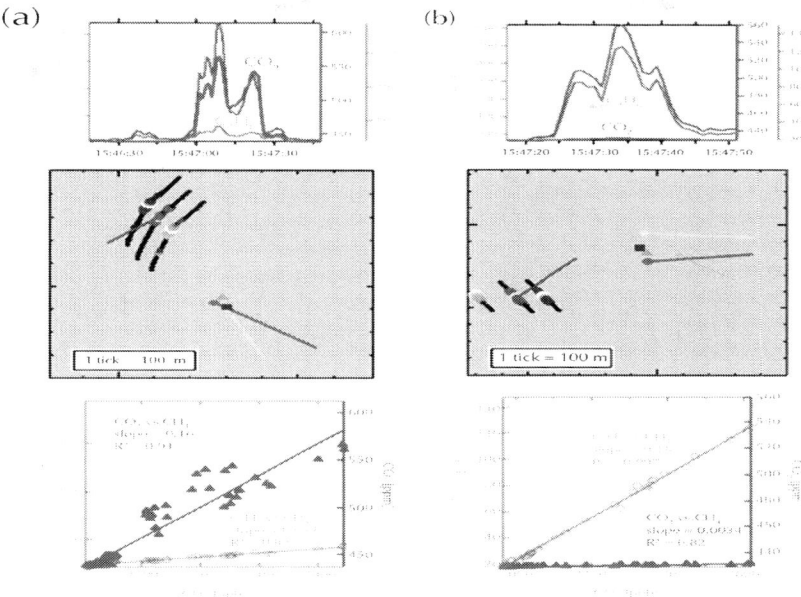

Figure 8: Example of differing CO_2 plume profiles as a function of gas play. (a) Emissions from a plant in a coal-gas region, with an amine scrubbing unit, showing significant CO2 emissions, and (b) emissions from a gathering facility with no treatment in a shale gas region.

Natural Gas Processing

Natural gas processing plants are large, complex facilities that remove unwanted compounds in the incoming gas stock (e.g. H_2S, CO_2, H_2O) and separate other high value compounds (i.e. natural gas liquids, as discussed below) from the gas to produce pipeline quality natural gas. Physically, processing plants often serve as the nexus between the gathering networks in the area and a transmission system working to serve 20 longer range transport. They are typically characterized by capacity throughputs of 3– 1500 million standard cubic feet per day (MMscfd). The types of equipment and the processes that are undertaken at a gas-processing plant depend on the composition of the gas in the region. Many plants utilize multiple processing "trains" to enable flexible operation. The equipment and steps in each train can vary depending again on the 25 region and the engineering decisions made by the operator of the plant (Kidnay et al., 2011). It should also be noted that not all natural gas in the US supply chain is processed.

Rather, in cases where natural gas composition does not contain substantial levels of natural gas liquids or $H_2S=CO_2$ (i.e. is dry and sweet), the natural gas flows directly from gathering facilities into transmission pipelines (and sometimes directly into distribution networks).

The initial process that is typically found at a gas-processing plant involves a continuation of the treatment types found in the gathering system of the region. At some 5 facilities, the initial product will be a first cut at collecting natural-gas condensate, which is typically comprised of functionalized hydrocarbons above C5, using an inlet separator (if they have not been collected further upstream in the gathering network). Water may also be removed using glycol dehydration. Other trace contaminants are often filtered using a series of molecular sieve apparatus that are staggered for effective continu10 ous regeneration. As discussed below, natural gas liquids (NGLs) are removed from the gas stream using either a cryogenic separation or separation based on solubility in lean oil (Kidnay et al., 2011). Additional details of this class of compounds and specific equipment used are discussed in the next section.

Due to the nature of the various processing steps and types of equipment found 15 at processing plants, as well as the somewhat larger geographic scale they typically occupy, there are typically multiple methane emission points, with various co-emitted compounds. On the surface, this type of source is a direct challenge to the tracer release methodology given the constraint for the controlled tracer release to be as close to the emission source as possible. The following examples and discussion describe 20 how these types of facility are quantified using the dual tracer methodology as well as using the nature of the co-emitted compounds to deduce the dominant emission sources.

The geographic scale of processing plants presents a challenge to the dual tracer flux ratio quantification given the constraints of wind direction and roadway access. Figure 9 depicts a pair of transects from a processing plant. Each transect was collected with the mobile lab maneuvering north to south. This is depicted by the rainbow bar in each of the two split time series (a) and (b) in the left hand panel and portrayed on the right hand panel with the relative distance (north vs.

east). In the case where the transect was captured at the facility fence line (a), we see relatively high spikes in plume mixing ratios with three different quantifiable E =M ratios. Note that the tracer release locations were relatively close to one another and this is reflected in the spatial coherence in both of the transects.

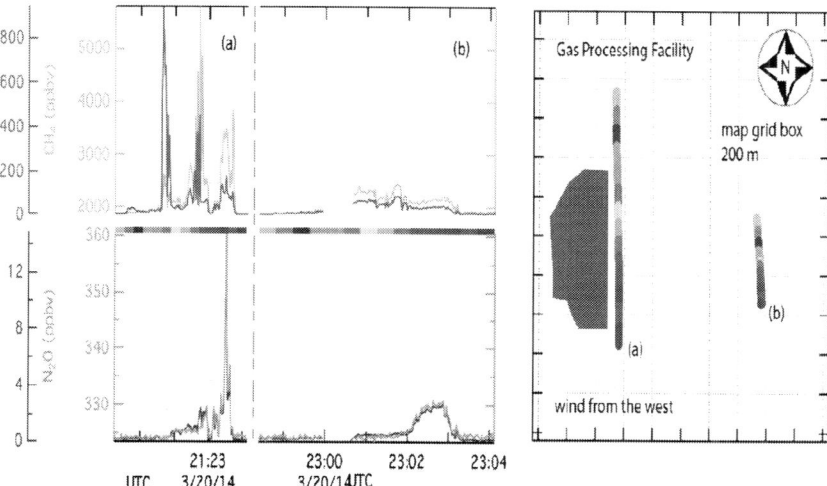

Figure 9: In the left hand panel, the time series for methane, ethane, nitrous oxide and acetylene are depicted for two transects, (a) and (b). In the right hand panel, the geographic location is portrayed for the processing plant (grey) and the two transects (a) and (b). See text for additional discussion.

In the case of the more distant (~ 1.2 km) transect, the mixing ratios of ethane and 5 methane are significantly less spiked. Careful analysis of the time and space dependence of the E=M ratio suggests that even at this distance, the ratio in the northern sector of the facility is different than that in middle and southern sections. This observation is corroborated anecdotally by the physical location of the liquids storage and NG-transmission hardware onsite. In this facility the recompression of pipeline grade 10 natural gas takes place in the southern third of the facility. This corresponds to the lowest E=M ratio (red-purple in the time series), but is a significant source of CH_4 emissions (~ 50 %) from the facility. The liquids storage and handling takes place at the northern section of the facility. The effective leak rate of methane is less than in other sections of the facility because the methane is at residual levels in the liquids 15 headspace. The E =M ratio in the green and yellow section of the time series is greater because this is where the NGL stock is being processed.

To quantify the FLER from processing facilities, frequently the dual-area analysis method is used. In the case of the close transects, the measured methane emission rates often exhibit substantial variance. The average of multiple close transects typically 20 was found to be comparable to values determined by more distant, better mixed plume intercepts, when such a comparison was available.

Natural Gas Liquids and Condensates

Natural gas liquids (NGL) is an umbrella term (EIA, 2013) for the many different chemicals and blends extracted in the liquid form from natural gas. Depending on the equipment available and the demand for the various products, the amount of processing of natural gas can vary greatly. At the lower end of the spectrum, the gas may undergo dehydration and just enough removal of C_2+ to meet pipeline specifications, such that no liquids condense at pipeline pressures. Removal of other impurities such as CO_2 and H_2S may also be required to meet pipeline specifications. At the highest end of the processing spectrum, cryogenic distillation will be employed to sequentially extract methane (demethanizer), ethane (deethanizer), propane, iso- and n-butane, and higher hydrocarbons. This processing can occur at a single facility, or can be performed in several 5 steps between different facilities. The net result is to separate the methane (and/or ethane) from other condensable compounds that may still be present in the feed stock after the various upstream treatments. The liquid *product* at this stage is referred to as "x" or "y" grade liquid depending on the cut temperature and ethane content in the liquid. In some of the processing plants in this study, this liquid stream is stored in this 10 state and shipped o_ site via an NGL pipeline or tanker truck. In other facilities studied, the liquid is further fractionated, sequentially removing ethane, then propane, then butane (Kidnay et al., 2011). Because of the low methane content within the liquid, this further processing of the NGL is not expected to significantly contribute to the FLER, but may play a role in the E=M ratio that is observed downwind.

Many of the facilities visited in this study were in so-called "ethane rejection" mode, meaning that distillation towers were operated at lower liquids recovery levels and purified ethane is treated as a byproduct of the C3+ extraction. As a byproduct, it frequently was re-injected into the natural gas stream. This occurs when there is less demand for purified ethane as a feedstock for ethylene, a process that occurs at an extremely 20 limited number of locations in the US.

As in the case of identifying condensate tank emissions, the E =M ratio can inform the attribution of a methane emission source to individual pieces of NGL equipment. A striking example is shown in Fig. 10. This facility has 5 compressors, 3 dehydrators, 5 condensate tanks, and desulfurization equipment. The nitrous oxide tracer (green 25 marker) was placed near the compressors and the acetylene tracer (blue marker) near the battery of condensate tanks. North-east of the acetylene tracer, above-ground piping marks the facility's inlet and outlet (natural gas) as well as a liquids pipeline carrying a mixture of ethane and propane produced at the facility. The E =M ratio for the mixed facility plume was 0.0576, while the ratio for the liquids pipeline and inlet/outlet region was 14.58, i.e. nearly entirely ethane. Therefore, this transect indicates that the pipeline is not a significant source of CH_4 emissions.

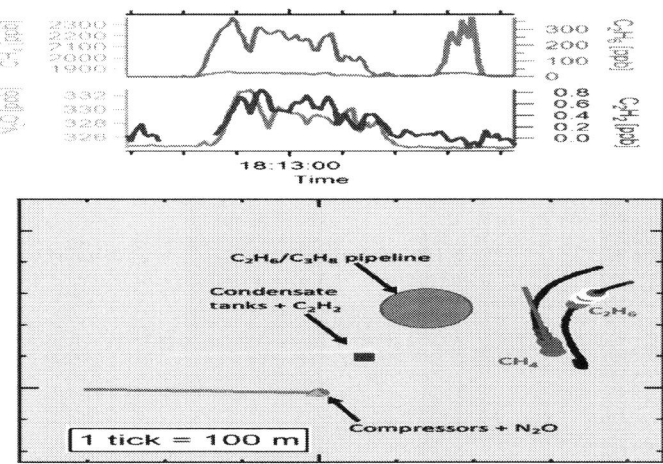

Figure 10: Downwind plume transect showing mixing ratio as a function of time (top) and a map (bottom). Tracer release locations are shown as a green

triangle (nitrous oxide) and a blue square (acetylene). The plume transect is colored by methane mixing ratio (black to yellow). Ethane mixing ratio is also shown with a geographic oset. Wind vectors (pink, red and yellow) point into the wind.

Comparison of C_2 Content with Operator Data

In this study, the E=M ratio serves several purposes: (i) confirmation that a plume is 5 from a target facility, (ii) elimination of plumes from neighboring facilities or biogenic sources, and (iii) distinguishing between different emission sources within a given facility. The quantification of a facility's methane emissions leverages (i) and (ii) above. Figure 11 shows a comparison between the measured E=M ratios at each facility and the operator-provided data on gas composition. Agreement is good overall, with a few 10 outliers. Also shown in the figure are 95% confidence limits on the measured E =M ratios. Large error bars in the facility average for E =M ratios are usually due to variations in the emission composition, since the error for any individual ratio measurement is low. The operator gas composition information was not always measured on the same day as the field testing. For gathering facilities, gas composition is periodically measured by gas sampling and subsequent third party analysis. For processing plants, gas composition data is typically acquired in real time at multiple locations at the facility. In either case, the gas composition exiting the gathering facility or processing plant may not always reflect the gas composition of the emission sources. This can be due to the E=M ratio changing as the gas moves through the facility, or from emissions from con20 densate/produced water tanks. This variety of equipment and processes at gathering facilities and processing plants explains much of the discrepancy between measured and operator E=M ratios, as compared to the transmission and storage study, where the composition of the gas does not vary during handling (Subramanian et al., 2014; Yacovitch et al., 2014). Table 3 outlines the minimum, median and maximum facility av25 erage E =M ratios divided by primary gas type. It should be noted that the classification by gas type is not rigid. That is, there may be multiple gas types other than the primary present at these facilities. The points in Fig. 11 are colored based on this gas classification. As noted above, coal bed methane facilities typically have the lowest E =M ratios. Conventional facilities sit somewhere in the middle, with the shale gas facilities split into several clusters. The shale gas is scattered about the plot, with some clustering associated with various geographic basins. The three main shale clusters observed in Fig. 11 (green points) correspond loosely to: the Denver (Denver-Julesburg), Permian 5 (Eagle Ford and Delaware), and Appalachian basins ($\sim$ 12–23%); the Anadarko (Mississippian Lime Gas play), Uinta (Natural Buttes) and Piceance basins ($\sim$ 4–6 %); and the Arkoma basin ($\sim$ 1 %). Other shale basins were also visited but the number of facilities for each of these basins is low.

Table 3: Measured E =M ratios as a function of gas type at gathering and processing facilities. Minimum, median and maximum average measured ratios are noted. Offshore gas is not included here due to the small number of o_shore facilities measured

Gas Type	Measured E / M ratio			
	min	median	max	Count
Coal Bed Methane	0.00	0.014	0.045	8
Coal Bed Methane and Conventional	0.0057	0.018	0.031	4
Shale	0.0055	0.051	0.24	64
Conventional	0.012	0.068	0.22	37

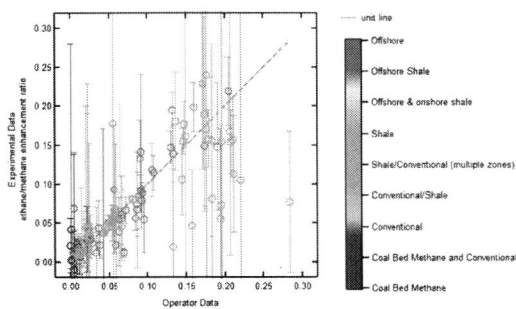

Figure 11: Comparison between measured ethane=methane ratio and operator data on gas composition. Error bars correspond to the 95% confidence

limits from the replicate experimental plumes. Points are also colored by the type of gas at each site. A line to guide the eye is drawn at a 1: 1 correspondence between measured and operator data.

CONCLUSIONS

Reported here is a detailed description of the measurement and analysis methods used during a field campaign to quantify methane emissions and emission sources from natural gas gathering and processing facilities. The campaign covered a wide range of geographic regions, basins, gas types and facilities. The measurement method used, dual tracer ratio, yielded facility-specific methane emission rates for 130 facilities. The field 15 measurements were complemented by onsite infrared imaging and equipment surveys. The analysis technique applied to the data allowed for accurate determination of the emission rates using multiple downwind plume categories. Overall emission profiles, quantified by measuring CH_4, C_2H_6, CO, CO_2, C_2H_2, and N_2O frequently afforded an understanding of the unique chemical signatures associated with various natural gas 20 gathering and processing equipment onsite. This paper provides a background and method description for additional work aimed at compiling the dataset (Mitchell et al., 2014) and developing an estimate (with uncertainty) for the total methane emissions from gathering and processing in the US (Marchese et al., 2014). *Acknowledgements.* This work was funded by the Environmental Defense Fund (EDF), as well as Access Midstream, Anadarko Petroleum Corporation, Hess Corporation, Southwestern Energy and Williams. These sponsors also provided important technical insight, facility data, and site access. Funding for EDF's methane research series is provided by Fiona 5 and Stan Druckenmiller, Heising-Simons Foundation, Bill and Susan Oberndorf, Betsy and Sam Reeves, Robertson Foundation, Alfred P. Sloan Foundation, TomKat Charitable Trust, and the Walton Family Foundation. Although not a financial sponsor, DCP Midstream provided technical insight and site access to a processing plant in the DJ Basin. Ramon Alvarez and Drew Nelson of EDF provided valuable technical and logistical support and guidance. We also 10 acknowledge David T. Allen, Garvin Heath, Michael Levi and James McCarthy of the Science Advisory Panel for providing important advice, guidance and input. The authors would like to thank John Nowak, Kenji Lizardo,

Jason Curry, Ryan McGovern, Mike Agnese, Bill Brooks, Xavier Cabral, and Katie Taylor for their contributions in the field during this project.

REFERENCES

1. Allen, D. T., Torres, V. M., Thomas, J., Sullivan, D. W., Harrison, M., Hendler, A., Herndon, S. C., Kolb, C. E., Fraser, M. P., Hill, A. D., Lamb, B. K., Miskimins, J., Sawyer, R. F., and Seinfeld, J. H.: Measurements of methane emissions at natural gas production sites in the United States, P. Natl. Acad. Sci. USA, 110, 17768–17773, 2013.

2. Beychok, M. R.: Fundamentals Of Stack Gas Dispersion, Newport Beach, CA, 2005. 20 Bottoms, R. R.: Separating Acid Gases, US Patent 1783901, 1930.

3. Bullock, A. and Nettles, R.: Remote Sensing VOC Project, Texas Commission on Environmental Quality, Texas, USA, 2014.

4. Crosson, E. R.: A cavity ring-down analyzer for measuring atmospheric levels of methane, carbon dioxide, and water vapor, Appl. Phys. B, 92, 403–408, 2008.

5. Czepiel, P. M., Mosher, B., Harriss, R. C., Shorter, J. H., McManus, J. B., Kolb, C. E., Allwine, E., and Lamb, B. K.: Landfill methane emissions measured by enclosure and atmospheric tracer methods, J. Geophys. Res., 101, 16711–16719, 1996.

6. Draxler, R. R. and Hess, G. D.: Description of the HYSPLIT_4 Modeling System, NOAA Air Resources Laboratory, Silver Spring, MDERL ARL-224, 24 pp., 1997.

7. Draxler, R. R. and Hess, G. D.: An Overview of the HYSPLIT_4 modeling system of trajectories, dispersion, and deposition, Aust. Meteorol. Mag., 47, 295–308, 1998.

8. EIA: Natural Gas Processing: The Crucial Link Between Natural Gas Production and its Transportation to Market, Energy Information Administration, 2006.

9. EIA: EIA's Proposed Definitions for Natural Gas Liquids, U.S. Energy Information Administration, Washington, DC, 5 pp., 2013.

10. EPA: Replacing Wet Seals with Dry Seals in Centrifugal Compressors, Natural Gas STAR Pro10 gram, 2006.

11.   EPA: National Greenhouse Gas Emissions Data, available at: http://epa.gov/climatechange/ ghgemissions/usinventoryreport. html (last access: September 2014), 2014a.

12.   EPA: Annual Energy Outlook 2014 with Projections to 2040, EPA Energy Information Administration, 2014b.

13.   Goetz, J. D., Floerchinger, C., Fortner, E. C., Wormhoudt, J., Massoli, P., Knighton, W. B., Herndon, S. C., Kolb, C. E., Knipping, E., Shaw, S. L., and DeCarlo, P. F.: Atmospheric Emission Characterization of Marcellus Shale Natural Gas Development, Environ. Sci. Technol., submitted, 2014.

14.   Harrison, M. R., Galloway, K. E., Hendler, A., Shires, T. M., Allen, D. D., Foss, D. M., Thomas, J., 20 and Spinhirne, J.: Natural Gas Industry Methane Emission Factor Improvement Study, Final Report, Cooperative Agreement No. XA-83376101, 2011.

15.   Herndon, S. C., Jayne, J. T., Zahniser, M. S., Worsnop, D. R., Knighton, B., Alwine, E., Lamb, B. K., Zavala, M., Nelson, D. D., McManus, J. B., Shorter, J. H., Canagaratna, M. R., Onasch, T. B., and Kolb, C. E.: Characterization of urban pollutant emission fluxes and ambi25 ent concentration distributions using a mobile laboratory with rapid response instrumentation, Faraday Discuss., 130, 327–339, 2005.

16.   Jiménez, R., Herndon, S., Shorter, J. H., Nelson, D. D., McManus, J. B., and Zahniser, M. S.: Atmospheric trace gas measurements using a dual quantum-cascade laser mid-infrared absorption spectrometer, Proceedings of SPIE, 2005,

17.   Jumonville, J.: Tutorial on cryogenic turboexpanders, in: Proceedings of the 39th Turbomachinery Symposium, 2010.

18.   Karion, A., Sweeney, C., Pétron, G., Frost, G., Michael Hardesty, R., Kofler, J., Miller, B. R., Newberger, T., Wolter, S., Banta, R., Brewer, A., Dlugokencky, E., Lang, P., Montzka, S. A., Schnell, R., Tans, P., Trainer, M., Zamora, R., and Conley, S.: Methane emissions estimate from airborne measurements over a western United States natural gas field, Geophys. Res. Lett., 40, 4393–4397, doi:10.1002/grl.50811, 2013.

19.   Kidnay, A. J., Parrish, W. R., and McCartney, D. G.: Fundamentals of Natural Gas Processing, 5 2nd ed., CRC Press, 2011.

20. Lamb, B., Westberg, H., and Allwine, G.: Isoprene emission fluxes determined by an atmospheric tracer technique, Atmos. Environ., 20, 1–8, 1986.

21. Lamb, B., McManus, J. B., Shorter, J. H., Kolb, C. E., Mosher, B., Harriss, R. C., Allwine, E., Blaha, D., Howard, T., Guenther, A., Lott, R. A., Siverson, R., Westberg, H., and Zimmer10 man, P.: Development of atmospheric tracer methods to measure methane emissions from natural gas facilities and urban areas, Environ. Sci. Technol., 29, 1468–1479, 1995.

22. Lamb, B. K., Edburg, S. L., Ferrera, T. W., Howard, T., Harrison, M., Kolb, C. E., Townsend- Small, A., Dyck, W., Possolo, A., and Whetstone, J.: Direct measurements show decreasing methane emissions from natural gas local distribution systems in the United States, Environ. 15 Sci. Technol., submitted, 2014.

23. Marchese, A. J., Zimmerle, D., Vaughn, T. L., Martinez, D., Williams, L., Robinson, A. L., Mitchell, A. L., Subramanian, R., Tkacik, D. S., Roscioli, J. R., and Herndon, S. C.: Methane emissions from United States natural gas gathering and processing, in preparation, 2014.

24. McKain, K.: Framework to Quantify Methane Emissions from Natural Gas Leakage in Boston, 20 American Geophysical Union Fall Meeting, 2012.

25. McManus, J. B., Nelson, D. D., Shorter, J. H., Jiminez, R., Herndon, S., Saleska, S., and Zahniser, M. S.: A high precision pulsed QCL spectrometer for measurements of stable isotopes of carbon dioxide, J. Mod. Optic., 52, 2309–2321, 2005.

26. Mitchell, A. L., Tkacik, D. S., Roscioli, J. R., Herndon, S. C., Yacovitch, T. I., Martinez, D. M., 25 Vaughn, T. L., Sullivan, M., Floerchinger, C., Marchese, A., and Robinson, A. L.: Measurements of methane emissions from the United States natural gas gathering stations and processing plants: measurement results, Environ. Sci. Technol., submitted, 2014.

27. Mosher, B. W., Czepiel, P. M., Harriss, R. C., Shorter, J. H., Kolb, C. E., McManus, J. B., Allwine, E., and Lamb, B. K.: Methane emissions at nine landfill sites in the Northeastern United 30 States, Environ. Sci. Technol., 33, 2088–2094, 1999.

28. Nehrkorn, T., Eluszkiewicz, J., Wofsy, S. C., Lin, J. C., Gerbig, C., Longo, M., and Freitas, S.: CoupledWeather Research and

Forecasting-Stochastic Time-Inverted Lagrangian Transport (WRF-STILT) Model, Meteorol. Atmos. Phys., 107, 51–64, 2010. Pétron, G., Frost, G., Miller, B. R., Hirsch, A. I., Montzka, S. A., Karion, A., Trainer, M., Sweeney, C., Andrews, A. E., Miller, L., Kofler, J., Bar-Ilan, A., Dlugokencky, E. J., Patrick, L., Moore, C. T., Ryerson, T. B., Siso, C., Kolodzey, W., Lang, P. M., Conway, T., Novelli, P., Masarie, K., Hall, B., Guenther, D., Kitzis, D., Miller, J., Welsh, D., Wolfe, D., Ne_, W., and Tans, 5 P.: Hydrocarbon emissions characterization in the Colorado Front Range: a pilot study, J. Geophys. Res.-Atmos., 117, D04304, doi:10.1029/2011JD016360, 2012.

29. Pétron, G., Frost, G. J., Trainer, M. K., Miller, B. R., Dlugokencky, E. J., and Tans, P.: Reply to comment on: "Hydrocarbon emissions characterization in the Colorado Front Range: a pilot study" by Levi, M. A., J. Geophys. Res.-Atmos., 118, 236–242, doi:10.1029/2012JD018487, 10 2013.

30. Rella, C., Crosson, E., Thoma, E., Hater, G., Merrill, R., Tan, S., and Green, R.: An Acetylene Tracer-Based Approach to Quantifying Methane Emissions from Distributed Sources Using Wavelength-Scanned Cavity Ring-Down Spectroscopy, American Geological Union Fall Meeting, 2009.

31. Rochelle, G. T.: Amine scrubbing for CO2 capture, Science, 325, 1652–1654, 2009. Shorter, J. H., McManus, J. B., and Kolb, C. E.: Collection of leakage statistics in the natural gas system by tracer methods, Environ. Sci. Technol., 31, 2012–2019, 1997.

32. Subramanian, R., Williams, L., Vaughn, T. L., Zimmerle, D., Roscioli, J. R., Herndon, S. C., Yacovitch, T. I., Floerchinger, C., Tkacik, D. S., Mitchell, A. L., Sullivan, M., and Robinson, A. L.: 20 Methane Emissions from Natural Gas Compressor Stations in the Transmission and Storage Sector: Measurements and Comparisons with the EPA Greenhouse Gas Reporting Program Protocol, in preparation, 2014.

33. Torres, V., Herndon, S. C., Kodesh, Z., and Allen, D. T.: Industrial flare performance at low flow conditions. 1. Study overview, Ind. Eng. Chem. Res., 51, 12559–12568, 2012.

34. Whiticar, M. J.: Correlation of natural gases with their sources, in: M 60: The Petroleum System- From Source to Trap, edited by: Magoon, L. B. and Dow, W. G., 261–283, 1994.

35. Yacovitch, T. I., Herndon, S. C., Roscioli, J. R., Floerchinger, C., McGovern, R. M., Agnese, M., Petron, G., Kofler, J., Sweeney, C., Karion, A., Conley, S. A., Kort, E. A., Nähle, L., Fischer, M., Hildebrandt, L., Koeth, J., McManus, B. J., Nelson, D. D., Zahniser, M., and Kolb, C. E.: 30 Demonstration of an ethane spectrometer for methane source identification, Environ. Sci. Technol., 48, 8028–8034, 2014.

36. Zavala-Araiza, D., Sullivan, D. W., and Allen, D. T.: Atmospheric hydrocarbon emissions and concentrations in the Barnett Shale natural gas production region, Environ. Sci. Technol., 48, 5314–5321, 2014.

37. Zimmerle, D., Williams, L., Vaughn, T., Subramanian, R., Duggan, J., Willson, B., Opsomer, J., 5 and Robinson, A.: Estimate of methane emissions from the natural gas transmission and storage system in the United States, in preparation, 2014.

Chapter 5

Electric Motor Drive for Natural Gas Compression in Pipeline: Techno-economic Analysis

A. Nasir[1], P. Pilidis[2], S. Ogaji[3], H. T. Abdulkarim[4], and A. El-Suleiman[5]

[1]PhD Researcher, Dept. of Power and Propulsion, Cranfield University, Cranfield, United Kingdom

[2]Professor and Head Dept. of Power and Propulsion, Cranfield University, Cranfield, United Kingdom

[3]Engineering Consultant and Technical Adviser, Ministry of Power, Abuja

[4]Lecturer, Dept. of Electrical/Electronic Technology, College of Education, Minna, Nigeria

[5]PhD Researcher, Dept. of Power and Propulsion, Cranfield University, Cranfield, United Kingdom

ABSTRACT

The transportation of natural gas requires more compressors than any other sector of the oil and gas market due to the long distances the gas must travel between gathering, processing and distribution sites. Electric drive is one of the major prime movers used for these compressor stations in order to maintain the gas flow and to deliver at a set pressure. To maintain a minimum transportation cost, a techno-economic module which will guide the drive and pipe selection is important. This paper presents a techno-economic tool which can be used to rapidly assess the profitability or otherwise of a natural gas pipeline project employing an electric motor drive as a prime mover. The techno economic tool is made of modules which are integrated together by a FORTRAN code called wrapper. This controls the running of each of the modules. The modules which make up the techno-economic tool are pipeline and compressor station module, electric motor and economic modules. As a case study for this analysis, a 24 inch 512km pipeline with a natural gas throughput of 4.54 million cubic meters per day (160.3 MMscfd) requiring a drive power of 34 MW was employed. From the results and analysis, for a throughput of 0.5 Mm^3/day, the transportation cost is $0127, $0.192 and $0.249 for pipe sizes of 304.8 mm, 609.6 mm and 1219.2 mm respectively. The results presented also shows that the economic pipe size for a 4.5 million cubic meter per day of natural gas is 609.6 mm (NPS 24) with transportation cost of $0.043 which is equivalent to $1.13 per GJ.

INTRODUCTION

Increase in world's population and advancement in technology has led to tremendous rise in energy demand. Natural gas is the most used fossil fuel and will remain the preferred fuel to meet the ever growing energy demand for the foreseeable future. It is expected that global consumption of gas will double by 2030 [1]. For natural gas to meet the energy demand, it is required to be transported from the production region through processing plant to the customers. This process certainly requires compressor stations along a pipeline which requires a prime mover in order to keep the gas flowing and maintain a set delivery pressure.

Other than gas turbine, another viable prime mover option for natural gas compressor is the electric drive. Natural gas pipeline deregulation is causing pipeline operators to evaluate electric drives **as** an alternative to gas driven equipment [2]. In the last decade, electric motor driven compression has become more common in the natural gas industry. Many of the components of an electric motor drive system have undergone technological changes to meet the needs of gas compressor applications. The evolving of variable frequency drives (VFD), variable speed drives (VSD), and motors with advance bearing technologies has provided pipeline compression a more efficient drive system, with larger and more flexible operating envelopes [3]. There is better energy conversion efficiency in electric motor than in gas turbine, with over 95% of the electrical energy coming in being converted into mechanical energy going out. This high electric motor efficiency can further be improved with VSD which modulate motor output by varying the speed of the motor itself, rather than through the use of control valves. The use of electric drive for pipeline compressors has come to stay because of its inherent advantage over some other drive option. These advantages include the lower maintenance cost and very importantly lack of on-site emission. This paper presents the techno-economic evaluation of electric drive for pipeline compressor using a model developed for rapidly assessing the profitability or otherwise of the system.

TYPES OF ELECTRIC MOTOR AND DRIVE TRAIN CONFIGURATION

Electric motors can be either alternating current (A.C.) or direct current (D.C.) motors. Figure 1 shows the classification of electric motors. Amongst the numerous types of electric motor, induction motors are the most commonly employed as prime movers for various industrial equipment. Theirpopularity is due to their simple design, ruggedness, low cost and easy maintenance and can bedirectly connected to an AC power source.The induction motor works by inducing current in the rotor through the small air gap between the stator and the rotor.

The stator current generates a rotating magnetic field in the air gap between the stator and rotor. The interaction of the induced rotor current with the rotating magnetic field generates a torque on the rotor. The synchronous speed which is the rate of rotation of the rotating magnetic field created by the stator is given by the equation (1);

$$n_s = \frac{60 \times f}{p}$$

(1)

Where f is the frequency of the AC supply current in Hz and p is the number of magnetic pole pairs per phase.

A major characteristic of the induction motor is the presence of slip (S) which is the difference between the rotating speed of the magnetic field (synchronous speed) and the rotating speed of the rotor. This determines the motor's torque and can be calculated from equation (2).

$$\%S = \frac{n_s - n_r}{n_s}$$

(2)

Where n_s is stator electrical speed and n_r is rotor mechanical speed.

The induction motor speed can be controlled to suit natural gas compressors requiring variable speed operation. The speed control can be achieved by varying input voltage, varying input frequency, changing the winding pole number or by varying input frequency and voltage together. Compressor variable speed requirement can also be met by a constant speed motor, utilizing a variable speed gearbox. The rotor speed, which is the speed delivered to the pipeline compressor if it is a direct coupling or to a gear arrangement if this exists, can be obtained from equation (3) with the knowledge of the percentage slip and synchronous speed [4].

$$n_r = (1 - s)n_s$$

(3)

Module Development and Integration

The procedure follows the development of a techno-economic module (TEM) architecture which can be employed to assess quickly the economic and technical profitability of a pipeline system using electric motor as a prime mover. Figure 2 shows the architecture for TEM for pipeline. The electric motor, pipeline, compressor station and economic modules were developed in FORTRAN codes. All the modules were integrated by a code called wrapper which controls the running of all the modules with results written into separate files. The development of the pipeline and compressor station modules have been described in earlier publication by the author, this relates to the use of gas turbine as prime mover [5]. The economic module developed considers the cost components of electric motor drive and the economic appraisal was done using net present value methodology. The electric motor module takes into cognisance the speed-torque relationship of electric motor vis-à-vis the operating speed of the compressor.

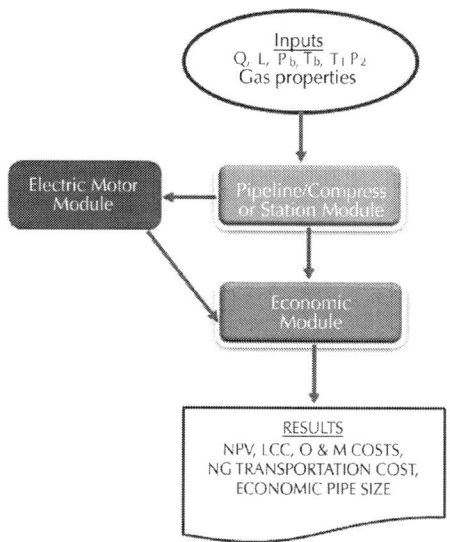

Figure 1: Techno-economic Module Architecture.

Development of Equivalent Circuit of an induction Motor

In developing an equivalent circuit of an induction motor, the similarity between a transformer and induction motor is considered. The primary of the transformer is similar to the stator of the induction motor and the rotor corresponds to the secondary of the transformer. It follows from this analogy that the stator and the rotor have their own respective resistances and leakage reactance. A magnetizing reactance exists because the rotor and the stator are magnetically coupled. The air gap in an induction makes the magnetic circuit relatively poor, thus the corresponding magnetizing reactance will be relatively smaller than that of transformer. The hysteresis and eddy current losses in an induction motor can be represented by a shunt resistance, as was done for the transformer.

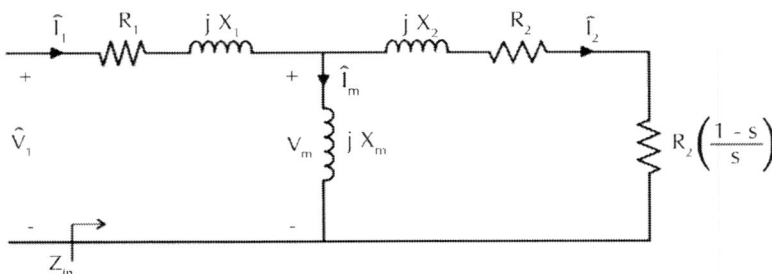

Figure 2: Equivalent circuit of an induction motor.

Figure 3 shows the equivalent electric circuit of an induction motor which consists of the traditional five parameters (i.e. stator resistance R_1, stator leakage reactance X_1, magnetizing Reactance X_m, rotor leakage reactance X_2, and rotor resistance R_2. Once the slip is calculated, the input impedance can be obtained from equation 4.

$$Z_{in} = R_1 + jX_1 + jX_m \| \left(\frac{R_2}{s} + jX_2 \right)$$

(4)

This can be further expressed mathematically as $Z_{in=}$

$$R_1 + jX_1 + \frac{jX_m \left(\frac{R_2}{s} + jX_2 \right)}{\frac{R_2}{s} + j(X_m + X_2)}$$

(5)

With the knowledge of the motor source voltage, V_1, the stator current can be computed using equation 6.

$$\hat{I}_1 = \frac{\widehat{V}_1}{Z_{in}}$$

(6)

$$\hat{I}_2 = \left(\frac{jX_m}{\frac{R_2}{s} + jX_2 + jX_m} \right) \hat{I}_1$$

(7)

The rotor current is flowing through the term $R_2/2$, which may be represented as the series of combination of a pure resistance R_2 and a back-emf term $R_2 \left(\frac{1-s}{s} \right)$. The mechanical torque can then be computed as the power into the backemf term divided by the mechanical speed. This results in

$$\tau_m = \frac{3I_2^2 R_2}{\omega_m} \left(\frac{1-s}{s} \right)$$

(8)

Figure 4 shows the percentage torque and synchrous speed characteristic for an induction motor.

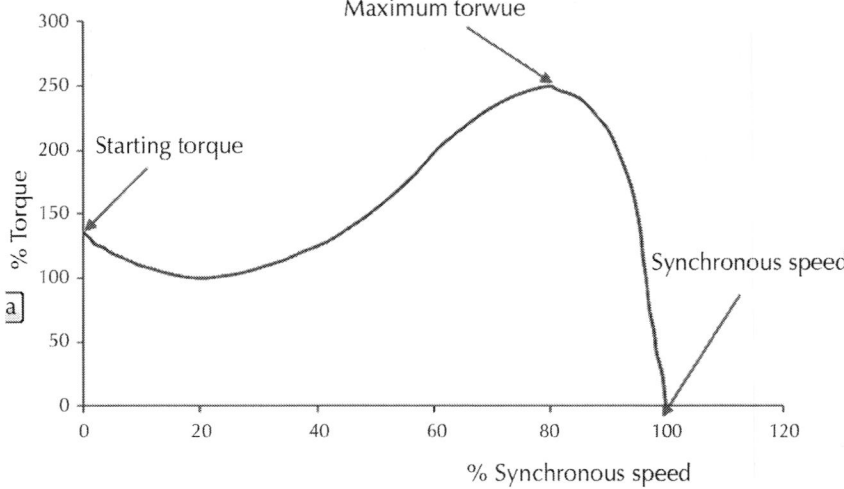

Figure 3: %Torque- % Synchronous speed curve of induction motor.

The torque versus speed relationship for the induction motor must be analysed carefully to ensure that all compressor required operating points may be met. The torque produced by an induction motor is a function of shaft power and the shaft speed, where the torque reduces with speed for constant power. This can be expressed as

$$\tau = 9.5493 \, \frac{P_m}{n_r}$$

(9)

Motor Life Cycle Cost (LCC)

Life cycle cost is the systematic economic consideration of all whole life costs and benefits of the motor over a period of analysis or expected motor life while fulfilling the performance requirements. This analysis is recommended to assess the large cost items in the motor installation and operation project. LCC is the capital cost (purchase and installation), plus maintenance and operation costs (based on energy prices) over its life time. This computation was done bearing in mind the life expectancy of the motor.

$$Electricity\ cost = \left(\frac{Motor\ power\ rating}{conversion\ efficiency} + P_{Loss}\right) \times electricity\ tariff$$

(10)

$$Transmission\ loss, P_{Loss} = \frac{P_T^2}{V^2}\left(\frac{\rho_{wire} \times L_{wire}}{A_{wire}}\right)$$

(11)

$$LCC = \sum_1^{n\ years} (Electricity\ cost + capital\ cost + O\&M\ cost)$$

(12)

RESULTS AND DISCUSSION

The results obtain from the simulation of the integrated TE modules are presented in this section. This result shows the economic appraisal of using electric motor drive as prime movers for pipeline compression. The results from the pipeline and compression station modules have been presented by the author in [5].

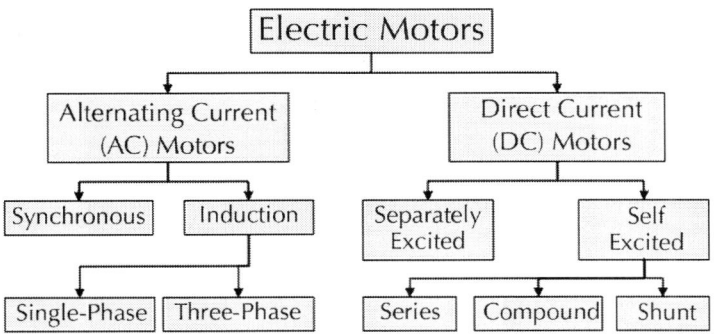

Figure 4

Effects of Throughput on the Operating Cost

Figure 5 presents the effect of throughput on the operating cost for electric motor drive option. The operating cost increased from $0.0615 billion to $0.623 billion as the throughput increased from 0.5 Mm³/day to 2.5 Mm³/day for a pipe size of 304.8 mm and electricity tariff of $0.05/kWh. This amounts to a difference of $0.562 billion. For 609.6 mm pipe size the operating cost increased from $0.0472 billion to $0.285 billion as the throughput increased from 0.5 Mm³/day to 2.5 Mm³/day which amounts to a difference of $0.237 billion. The rise in operating cost as a result of increase in throughput reduces with increase in pipe size. The sensitivity shows that the higher the electricity tariff, the higher the operating cost. Pipe size 1219.2 mm is seen to have the least rise in operating cost as drive power is minimal with flow through it. Figure 5 also shows that the operating cost tends to zero as the throughput reduces to zero.

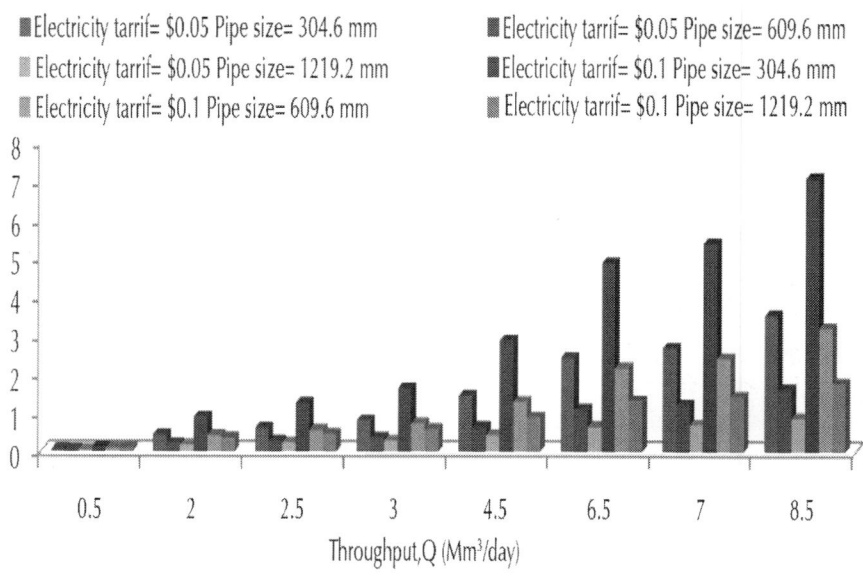

Figure 5: Operating cost variation with throughput for different pipe size.

Effect of Pipe Size on Operating and Pipe Material Cost

Figure 6 presents a comparative analysis of the effect of pipe size on pipe cost and operating cost.

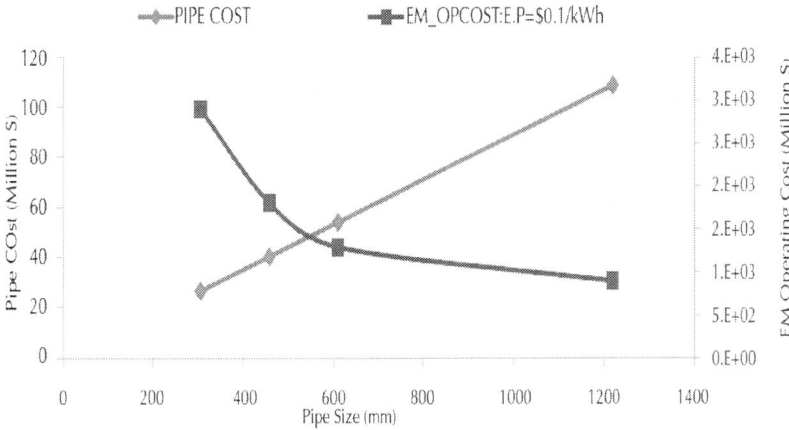

Figure 6: EM Operating cost & Pipe cost against with Pipe diameter for varying electricity price.

The pipe material cost for the usage of 609.6 mm size is $53.9 million and $26.5 million for a pipe size of 304.8 mm. This gives a savings of $27.4 million in material cost. On the other hand the operating cost for an estimated electricity tariff of $0.05/kWh is $641.9 million using a pipe size of 609.6 mm and for a pipe size of 304.8 mm and the same electricity tariff, the operating cost is $1.4 billion. This amounts to an increase of $804 million. It shows that although under-sizing of pipe apart from technical issues has negative economic impact on the natural gas pipeline project. The saving in pipe cost is only about 3.4 % of the increase in the operating cost.

Effect of Throughput on NPV

Figure 7 presents the effect of throughput on the Net Present Value for the Electric Motor for varying pipe sizes. For a throughput of 4.5 Mm³/day through a 609.6 mm pipe size the NPV is $2.7 billion and for a throughput of 7.0 Mm³/day, the NPV is $4.2 billion. This is an increase of about 55 % in NPV. But for a 4.5 Mm³/day through a 304.8 mm pipe size, the NPV is $2.6 billion and $4.0 billion for a throughput of 7.0 Mm³/day. This gives a percentage increase of 53.8%.

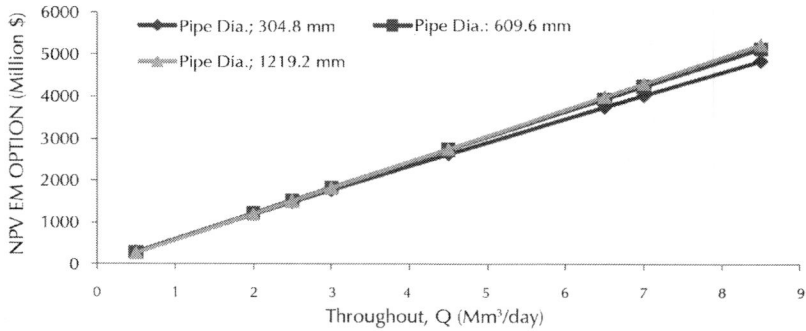

Figure 7: NPV Variation with throughput for different pipe sizes.

The increase in throughput undoubtedly increases the compressor drive power required, as shown in Figure 8. This rise in drive power will also cause an increase in the capital and operating costs.

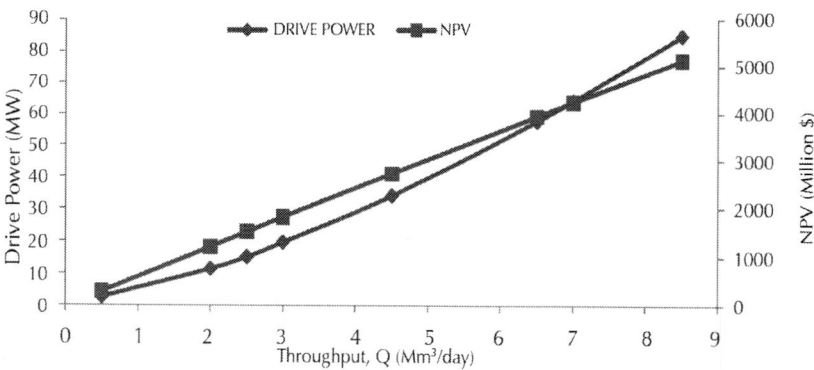

Figure 8: Drive power and NPV against throughput.

Figure 8 shows the trend of variation of drive power and NPV as the throughput changes. The increase in NPV, which puts together all the cost components and cash flows over the project life, despite the increase in capital and operating costs, further confirms the possession of economies of scale by pipeline transportation systems [6].

Effect of Throughput on Natural Gas Transportation Cost

Figure 9 presents the effect of throughput on gas transportation cost for varying pipe sizes using electric motor as driver. For a throughput of 0.5 Mm3/day, the transportation cost is $0.127, $0.192 and $0.249 for pipe sizes of 304.8 mm, 609.6 mm and 1219.2 mm respectively. It is seen that 304.8 mm presents the lowest cost at an electricity tariff of $0.05/kWh. This amount is equivalent to $3.34 / GJ, which is higher than what is expected of pipeline transportation cost and this suggest that 0.5 million cubic meter per day is not economical to be transported over long interstate pipelines.

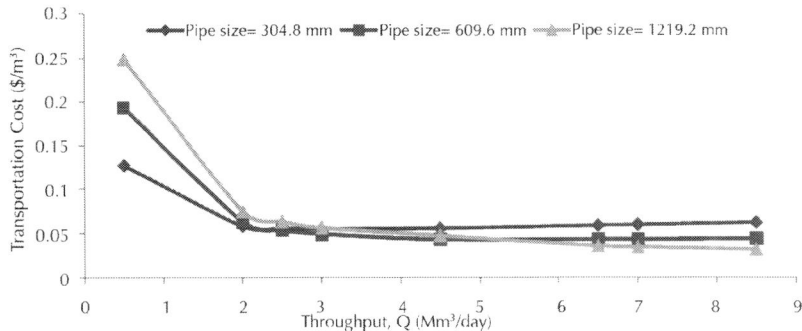

Figure 9: Gas Transportation cost.

An increase in throughput from 0.5 Mm³/day to 2.0 Mm³/day gave rise to a sharp drop in the gas transportation cost across all the pipe sizes and 609.6 mm pipe size has the least transportation cost of $0.054 although that is not the optimum point for this pipe size. Beyond 2.5 Mm³/day, an increase in transportation cost is noted for pipe size 304.8 mm. This implies that the optimum throughput for 304.8 mm is 2.5 Mm³/day which yields $0.055 transportation cost. For a 609.6 mm pipe size and electricity tariff of $0.05/kWh, the transportation cost rise as the throughput goes beyond 6.5 Mm³/day. For 1219.2 mm pipe size the transportation cost continues to drop all through the throughput studied for the studied electricity tariffs. It is believed that there will be an increase in the transportation cost at a point as the throughput continues to rise. This point will be the optimum throughput for 1219.2

mm pipe size. At a throughput of 8.5 Mm³/day, 1219.2 mm presents the lowest gas transportation cost of $0.032.

CONCLUSIONS

This paper presents the development of a techno-economic tool which can be used to rapidly assess the profitability or otherwise of a natural gas pipeline employing electric motor drive as prime mover for pipeline compressors. The results could be used to guide the selection of economic pipe size which gives the minimum investment and operating cost for the pipeline system and consequently minimum natural gas transportation cost. The transportation cost for a 0.5 million cubic meter per day of natural gas through a pipe size of 304.8 mm (12 inch) is $0.127/m³ which is equivalent to $3.34 /GJ. Although this is the minimum for this throughput, it is far higher than the transportation cost of about $1.4 /GJ found in confidential operators reports. This therefore, indicates that the transportation of natural gas of throughput of 0.5 Mm³/day or below over long distance pipeline is uneconomical.

ACKNOWLEDGMENT

The authors would like to thank the Petroleum Technology Development Fund (PTDF), Nigeria and the Federal University of Technology, Minna, Nigeria for the Scholarship and Study Fellowship Awards respectively.

REFERENCES

1.   Riva, A., D'Angelosante, S., andTrebeschi, C. "Natural Gas and The Environmental Results of Life Cycle Assessment", *Energy,* vol. 31, no. 1 SPEC. ISS, pp138-148, 2006.

2.   Oliver, J. A.., Corona, C., and Poteet, D. "High-Speed, High-Horsepower Electric Motors For Pipeline Compressors: Available ASD Technology", *IEEE Transactions on Energy Conversion,* Vol. 10, No. 3, pp470-476, 1995.

3.   Rama, J. C., and Gieseche, A. "High-speed Electric Drives: Technology and Opportunity", *IEEE Industry Applications Magazine,* Vol. 3, No. 5, pp48-55, 1997.

4.   Nored, M. G., Hollingsworth, J. R. and Brun, K. *"Application Guideline for Electric Motor Drive Equipment for Natural Gas Compressors"* , Gas Machinery Research Council, Southwest Research Institute, US, pp48-55, 2009.

5.   Nasir, A., Pilidis, P., Ogaji, S., and Mohamed, W. "Some Economic Implications of Deploying Gas Turbine in Natural Gas Pipeline Networks," *International Journal of Engineering and Technology,* Vol. 5, No. 1, pp141-145, 2013. Available online at http://www.ijetch.org/show-47-400-1.html.

6.   Subero, G, Sun, K., Deshpande, A. McLaughlin, J., and Economides, M. "A Comparative Study of Sea-Going Natural Gas Transport"*SPE Annual Technical Conference and Exhibition,* 26-29, September 2004, Houston, Texas, SPE, USA, pp1-7, 2004.

Chapter 6

Estimation of Regional Air-Quality Damages from Marcellus Shale Natural Gas Extraction in Pennsylvania

Aviva Litovitz[1], Aimee Curtright[2], Shmuel Abramzon[1], Nicholas Burger[3], and Constantine Samaras[2]

[1]RAND Corporation, 1776 Main Street, Santa Monica, CA 90407, USA
[2]RAND Corporation, 4570 Fifth Avenue, Pittsburgh, PA 15213, USA
[3]RAND Corporation, 1200 South Hayes Street, Arlington, VA 22202, USA

ABSTRACT

This letter provides a first-order estimate of conventional air pollutant emissions, and the monetary value of the associated environmental and health damages, from the extraction of unconventional shale gas in Pennsylvania. Region-wide estimated damages ranged from $7.2 to $32 million dollars for 2011. The emissions from Pennsylvania shale gas extraction represented only a few per cent of total statewide emissions, and the resulting statewide damages were less than those estimated for each of the state's largest coal-based power plants. On the other hand, in counties where activities are concentrated, NOx emissions from all shale gas activities were 20–40 times higher than allowable for a single minor source, despite the fact that individual new gas industry facilities generally fall below the major source threshold for NOx. Most emissions are related to ongoing activities, i.e., gas production and compression, which can be expected to persist beyond initial development and which are largely unrelated to the unconventional nature of the resource. Regulatory agencies and the shale gas industry, in developing regulations and best practices, should consider air emissions from these long-term activities, especially if development occurs in more populated areas of the state where per-ton emissions damages are significantly higher.

INTRODUCTION

Recent technological innovations in natural gas extraction— namely the combined use of horizontal drilling and hydraulic fracturing—are enabling access to vast new natural gas resources contained in shale deposits across the United States (Kargbo et al 2010, Mooney 2011). The Marcellus Shale formation is the largest US shale gas deposit and has contributed significantly in recent years to increased US natural gas production (US DOE EIA 2012a, 2012b). The rapid development of this resource has been touted as both an economic boon (Considine et al 2011, Marcellus Shale Coalition 2012) and a potential environmental mistake for the region (PennEnvironment Research and Policy Center 2012). Environmental concerns often relate to risks to water resources (Ground Water Protection Council and ALL Consulting 2009, Mooney 2011). However, utilizing natural gas from shale deposits also produces air emissions of various types during extraction, transportation, and end use.

Increases in conventional air pollution may pose a threat to air-quality in shale gas extraction regions (Shogre 2011, Alvarez and Paranhos 2012, McKenzie et al 2012, Steinzor et al 2012). Such emissions can have direct physical impacts on health, infrastructure, agriculture and ecosystems. For example, short-term exposure to criteria pollutants such as sulfur dioxide .SO_2/ and nitrogen oxides .NOx/ has been linked to adverse respiratory effects. Exposure to fine particulate matter (PM) and ozone .O3/ may increase respiratory-related hospital admissions, emergency room visits, and premature death. The expanded use of natural gas could arguably reduce net emissions from the electricity sector if used in lieu of coal (US EPA 1999, NRC 2010)[4]. However, shale gas extraction activities such as diesel truck transport and natural gas processing at compressor stations could lead to increases in air pollution in regions where extraction occurs.

[4] Emissions relative to renewable technologies are generally estimated to be lower than those of natural gas, so using natural gas in lieu of renewables would increase emissions.

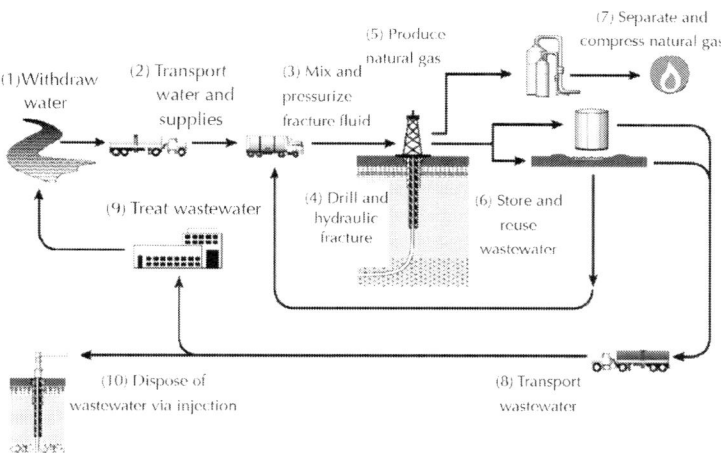

Figure 1: Major activities of shale gas extraction using horizontal drilling and hydraulic fracturing.

Life cycle greenhouse gas (GHG) emissions from shale gas are often assessed to be greater than conventional natural gas. However, most studies also indicate that expanded use of shale gas could lower net GHG emissions relative to coal-based electricity (Burnham et al 2011, Fulton et al 2011, Hultman et al 2011, Jiang et al 2011, Venkatesh et al 2011, Lu et al 2012, Skone et al 2012, Weber and Clavin 2012). Additionally, any GHG benefits from shale gas use are not localized to the region where extraction occurs. While GHGs are an important consideration, this letter focuses on conventional, non-GHG air pollution.

A recent GAO literature survey found evidence that extraction activities pose risks to air quality. While some studies indicated degraded air quality at specific shale gas extraction sites, the data necessary to quantify aggregate impacts were not available (US Government Accountability Office 2012). Pennsylvania recently mandated reporting on some emissions to the Pennsylvania Department of Environmental Protection (PA DEP), but this data collection has just begun (Pennsylvania Department of Environmental Protection 2011). This analysis provides initial, firstorder estimates of regional air emissions generated by Pennsylvania-based extraction activities[5] and associated ranges of potential regional monetized damages. These estimates must be considered in the context of other external costs and benefits of shale gas extraction and use, and should be refined as new data becomes available.

[5] This analysis does not specifically address acute damages resulting from short-term, high levels of exposure near well-sites but rather focuses on region-wide damages from a general degradation in air quality.

ESTIMATING LOCAL EMISSIONS AND REGIONAL DAMAGE FROM SHALE GAS EXTRACTION ACTIVITIES

The major stages of shale gas extraction considered here are depicted in figure 1, and emissions occur across many of them (NYS DEC 2011). This analysis includes emissions associated with four shale gas-related activities:

- Diesel and road dust emissions from trucks transporting water and equipment to the site, and wastewater away (stages 2 and 8 in figure 1);
- Emissions from well drilling and hydraulic fracturing, including diesel combustion (stage 4);

- Emissions from the production of natural gas, including on-site diesel combustion and fugitive emissions (stage 5); _ Combustion emissions from natural gas powered compressor stations (stage 7).

Table 1: Air emissions damages, localization of effects, and relevant pollutants of concern

Damage category	Damage location	Relevant emissions	Relevant stages	Inclusion in this analysis
Climate change	Local, regional, and global	GHG_s: CO_2, C_1-1_4, N_2O, O_3	Stages 2, 8: transport Stages 3, 4, 5: site activities Stage 7: processing	No GHGs included in this study
Air quality	Local and regional	VOCs, $NO_{x,,,}$ PM, SO_2, O_3, CO	Stages 2, 8: transport Stages 3, 4, 5: site activities Stage 6: wastewater storage and reuse	Development activities: (1) transport; (2) well drilling, hydraulic fracturing Ongoing activities: (3) production; (4) compressor stations
			Stage 7: processing	Pollutants: direct: VOCs, $NOx_{,,,}$ PM, SO_2; indirect O_3 via VOCs and $NO_{,,}$

We omit emissions from venting or flaring at well-sites (stages 4 and 5). The US EPA will prohibit this by 2015, requiring so-called 'green completions' which capture completions emissions rather than venting or flaring them (United States Environmental Protection Agency 2012), and many natural gas producers have already begun following this practice. Industry-reported emissions for venting are small relative to other sources; however flaring-emission estimates may have a more substantial impact[6].

For industry inventories that report venting, these emissions are less than 0.1% of VOCs from well drilling and hydraulic fracturing, as described in section 3.2. However, another source (NYS DEC 2011) estimates that total drilling, fracturing, and production PM emissions increase by 250% with flaring; NOx and VOCs increase by 120%. Assuming these increases, and that all wells flare completions emissions and all PM from flaring is PM2:5, additional damages are $5.7 million, or 18% of our high-bound total damage estimate.

Pollutants assessed were: volatile organic compounds (VOCs)[7]; NOx; PM10 (<10 _m); PM2:5 (<2.5 _m);[8] and SO2.[9] We focus on these due to their adverse impacts and regulatory status; accordingly, they often appear in facility permitting and emissions reporting, and all are included in the model used here to monetize damages. Table 1 summarizes air pollutants and extraction activities included in this analysis.

[7]The EPA defines VOCs to include organic compounds that undergo photochemical reactions in the atmosphere and does not include methane.

[8]PM10 typically includes all particles less than 10 _m and $PM_{2.5}$ all particles less than 2.5 _m. Thus PM_{10} includes $PM_{2.5}$ in most reporting. In industry reports, there is considerable uncertainty in PM size, and it is often assumed that all PM is smaller than 2.5 _m (i.e. PM_{10} D $PM_{2.5}$). $PM_{2.5}$ has much larger health effects than PM_{10}; this assumption therefore implies the maximal damage.

[9]In some cases sulfur oxides are reported as a mixture .SO_x/; in our damage calculations, we treat all SO_x as SO_2.

METHODS USED TO CALCULATE AIR POLLUTION DAMAGES

There is considerable uncertainty in emissions associated with shale gas development. This is due to a scarcity of emissions data and to actual differences in emissions caused by regional and site-specific variations in technology and processes[10]. The several estimation methods and data sources we use result in a wide range of estimates.

[10]In addition to differences in practices and technologies, well-specific variables that may influence emissions include length of well bore, number of fracturing stages, geographic location, and characteristics of the natural gas formation (e.g. wet or dry gas). For emissions reported by industry, we have little knowledge of estimation methodology.

For industry data used here, estimation methods are likely to have been used (e.g., an emissions factor approach) rather than empirical determinations. Such estimations often differ widely from empirical findings, especially for fugitive emissions (Chambers et al 2008, P´etron et al 2012), which are also subject to uncertainty (Levi 2012).

Our approach to estimating regional air pollution damages is modeled after another study of the external costs of energy production (NRC 2010). For each activity we have estimated emissions on a per well or per-unit-of-natural-gasproduced basis. Compressor station emissions are estimated per station. These emissions estimates allow us to obtain total statewide emissions, with resolution at the county-level, that we convert to statewide damages using the Air Pollution Emission Experiments and Policy (APEEP) model (Muller and Mendelsohn 2007, 2012). We first describe our approach for estimating emissions (sections 3.1–3.5) and then describe how these emissions were converted into monetary damages (section 3.6).

Estimates of Air Pollutant Emissions from Transport Trucks

Diesel trucks used to transport water and supplies to and from the well-site emit air pollutants. Our assumption of the total number of per well truck trips is based on the New York State Department of Environmental Conservation's (NYS DEC) 2011 Environmental Impact Statement (EIS) (NYS DEC 2011). The corresponding implied diesel emissions were estimated with emissions factors in the Greenhouse gases, Regulated Emissions, and Energy use in Transportation (GREET) model (US DOE Argonne National Labs (ANL) 2012) and in a recent National Research Council study (NRC 2010) for light-duty and heavy-duty vehicles, respectively. Truck traffic can also result in considerable road dust, which we include based on estimates in the NYS EIS. Additional details are provided in section S.1 (available at stacks. iop.org/ ERL/8/014017/mmedia). Table 2 provides the total per well transport emissions assumed.

Table 2: Range of assumed well-site development emissions in this analysis

Emissions activity	VOC	NO$_x$	PM$_{2.5}$	PK$_{10}$	SO$_x$
Total diesel and road dust development emissions (kg/well)	18-31	320-580	9.4-13232[a]	9.8-32[a]	0.47-0.79
Total well-site development emissions (kg/well)	150-170	3800-4600	87-130	87-130[b]	3.8-110

[a]PM$_{10}$ emissions were unavailable for heavy-duty trucks; in this case, it was assumed all diesel-related PM emissions were less than 2.5 _m. All road dust was also assumed less than 2.5 _m. Therefore aggregate PM$_{10}$ counts differ from PM$_{2:5}$ only in light-duty vehicle emissions; at the high end of our range, this difference is not significant.

[b]Industry reporting often assumes all PM emissions are less than 2.5 _m and so PM$_{10}$ counts are almost the same as PM2:5.

Table 3: Range of assumed well-site production emissions used in this analysis

Emissions activity	VOC	NO$_x$	PM$_{2:5}$	PM$_{10}$	SO$_x$
Total annual well-site production emissions per well (kg=well)	46–1200	520–660	9.9–50	9.9–50[a]	3.1–4.0

[a]Industry reporting often assumes all PM emissions are less than 2.5 μm and so PM$_{10}$ counts are here the same as PM$_{2.5}$.

Estimates of on-Site Air Pollutant Emissions from Well Construction

Well development generates emissions at the extraction site during well pad construction, drilling, and hydraulic fracturing. The range of well-site construction emissions used in this analysis were estimated using data reported by three major regional shale gas producers, including one set of emissions reported directly to us and two sets obtained through PA DEP as part of its Air Emissions Inventory for the Natural Gas Industry (PA DEP 2011, Pennsylvania Department of Environmental Protection 2011, Ramamurthy 2012). Details on these data sets and how they were used are provided in section S.2 (available at stacks.iop.org/ERL/ 8/014017/mmedia); final values used in this analysis are provided in table 2.

Estimates of Air Pollutant Emissions from Shale Gas Production

The ongoing production of shale gas also generates emissions. Data were obtained from two major regional operators and were used to establish low and high values of production emissions estimates, shown in table 3. Production emissions obtained for this analysis were less consistent between sources than construction emissions, although values are typically within an order of magnitude. In addition to differences between producers, this range may also reflect differences in the operators' reporting assumptions (see section S.3 available at stacks.iop.org/ERL/8/014017/mmedia).

Estimates of Air Pollutant Emissions from Compressor Stations

Emissions from compressor stations continue over the long term as natural gas is produced over the life of many wells. To estimate ranges of potential emissions from compressor stations, we reviewed permit applications for more than a dozen new facilities permitted in Pennsylvania in 2010 and 2011, as described in section S.4 (available at stacks.iop.org/ERL/8/014017/mmedia). We make use of the facility-wide potential-to-emit (PTE) emissions values, with ranges reflecting the lows and highs observed in our review. If most facilities are operating below capacity, they may fall at the lower end of the estimate; on the other hand, if they are not running optimally (e.g., frequent shut-downs and start-ups), the emissions could be even higher than indicated by PTE. Values in table 4 therefore represent a range of operating situations.

Aggregated Air Pollutant Emissions Estimates

We used per-facility emissions to estimate county-level and statewide emissions. We present total statewide aggregated emissions in table 5. These values represent the ranges of emissions in tables 2–4 applied to the following extraction activity assumptions for 2011: construction of 1741 wells; statewide shale gas production of nearly 1.1 trillion cubic feet; and operation of 200 recently developed compressor stations. County-level assumptions and values can be found in section S.5 (available at stacks.iop.org/ERL/8/014017/ mmedia).

Estimating Damages from Air Pollutant Emissions

For each of the four activities included in this analysis, emissions per well or per million cubic feet were used to estimate county-level emissions because damage per unit of pollution varies greatly with location. These county-level emissions were then converted into county-level annual damages using the APEEP model (Muller and Mendelsohn 2007, 2012). APEEP is an integrated assessment model that uses information derived from the air quality and epidemiological literature[11]. APEEP converts tons of 11 In considering the annual benefits of the Clean Air Act in 2000, APEEP gives a result of $48 billion compared to the US EPA's estimate of $71 billion. Muller and Mendelsohn argue that the US EPA work likely overstates benefits as it relies on air quality monitoring at sites that were out of attainment, sites likely to show greater changes in pollution levels than the country at large (2007). Note that in making the APEEP estimates, Muller and Mendelsohn use the US EPA's assumptions on value of a statistical life and concentration response function.

In considering the annual benefits of the Clean Air Act in 2000, APEEP gives a result of $48 billion compared to the US EPA's estimate of $71 billion. Muller and Mendelsohn argue that the US EPA work likely overstates benefits as it relies on air quality monitoring at sites that were out of attainment, sites likely to show greater changes in pollution levels than the country at large (2007). Note that in making the APEEP estimates, Muller and Mendelsohn use the US EPA's assumptions on value of a statistical life and concentration response function.

Table 4: Range of compressor station emissions estimates used in this analysis

Emissions activity	VOC	NO$_x$	PM$_{2:5}$	PM$_{10}$	SO$_x$
Total annual compressor station emissions (metric tons=facility)	11–45	46–90	1.4–5.5	1.4–5.5$_a$	0–1.7

Industry reporting often assumes all PM emissions are less than 2.5 _m and so PM$_{10}$ counts are here the same as PM$_{2:5}$.

Table 5: Statewide emissions estimates for shale gas development and production in 2011

	Statewide annual emissions (metric tons per year)				
Activities	VOC	NO$_x$	PM$_{23}$	PM$_{10}$	SO$_x$
Transport	31-54	550-1000	16-30	17-30	0.82-1.4
Well drilling and hydraulic fracturing	260-290	6600-8100	150-220	150-220	6.6-190
Production	71-1800	810-1000	15-78	15-78	4.8-6.2
Compressor stations	2200-8900	9300-18000	280-1100	280-1100	0-340

Total$_a$	2500-11 000	17 000- 28000	460- 1400	460- 1400	12- 540

These totals are reported to two significant figures, as are all intermediate emissions values in this document. The activity emissions may not exactly sum to the totals.

Table 6: Estimates of regional air pollution damages from Pennsylvania extraction activities in 2011

Activities	Timeframe	Total regional damage for 2011 ($2011)	Average per well or per MMCF damage ($2011)
Transport	Development	$320000810000	$180-$460 per well
Well drilling, fracturing	Development	$2 200000-$4700 000	$1 200-$2 700 per well
Production	Ongoing	$290000-$2 700000	$0.27-$2.60 per MMCF
Compressor stations	Ongoing	$4 400 000-$24 000000	$4.20-$23.00 per MMCF
(1)-(4) Aggregated	Both	$7 200000-$32 000000	NA

pollutant emitted into physical health and environmental damages, including mortality, morbidity, crop and timber loss, visibility, and effects on anthropogenic structures and natural ecosystems. The base APEEP model calculates age-specific health damages, recognizing that mortality risk and lost years of life will vary with age. Section S.6 (available at stacks.iop. org/ERL/8/014017/mmedia) provides additional details and damages for each county. The damage ranges given for each county are a result of the ranges in emissions estimates above; in addition, because of uncertainty in the size of PM, for activities 2–4 the low damage estimates assume none of the PM is PM2:5 and the high damage estimates assume that all PM is PM2:5. Complete damages by county and pollutant are found in tables S.11 and S.12 (available at stacks.iop.org/ ERL/8/014017/mmedia).

RESULTS

Regional Shale Extraction Air Pollutant Damage Estimates

The aggregated estimated regional damages associated with Pennsylvania shale gas extraction activities are shown in table 6. The total regional air-quality-related damages, at the level of development and production in Pennsylvania in 2011, ranged between $7.2 million and $32 million. These represent the sum of damages in all Pennsylvania counties. While per unit damages will vary greatly with location of the emissions, we also calculated the average per well or per MMCF damages. Some extraction activities occur in regions of Pennsylvania that influence the air quality of populated areas of other states; so while our estimates of emissions were confined to extraction activities in the state of Pennsylvania, these damages should be considered a regional impact, given that pollutants may cross the state border.

Development activities represent about a third or less of total extraction-related emissions (35–17% across the estimated range), whereas ongoing activities represent the majority of emissions (65–83% across the range). Compressor station activities alone represent 60–75% of all extractionassociated damages. Considering the relative importance of different pollutants, VOCs, NO_x, and $PM_{2.5}$ combined across all activities were responsible for 94% of total damages; across the range of estimates they contributed 34–33%, 59–20%, and 2–41%, respectively (shown by activity in table S.11 at stacks.iop.org/ERL/8/014017/mmedia).

Comparison of Air Pollutant Emissions and Damages to

other industrial sectors in Pennsylvania To assess the relative impact the shale gas industry might have on regional air quality, we compare the total emissions estimated for extraction activities in 2011 with net emissions from other major sectors of the Pennsylvania economy. We obtained data from the US EPA's 2008 National Emissions Inventory

(NEI) (US EPA 2008) and calculated statewide emissions (see section S.7 available at stacks.iop.org/ERL/ 8/014017/mmedia). These statewide totals are presented in table 7, along with the percentage of these total emissions that shale gas extraction activities in 2011 represent. Compared to total emissions from all industries reporting, the shale extraction industry in 2011 was producing relatively little conventional air pollution. Only NOx emissions are equivalent to more than 1% of statewide emissions across the entire estimated range.

Table 7: Magnitude of shale gas extraction industry relative to air pollutant emissions from other industrial sectors in Pennsylvania

Total sector or comparison	VOCs	NO_x		$PM_{2.5}$	PM_{10}	SO_x
Shale gas extraction industry in 2011. from table 5 (metric tons)	2500-11000	17 000-28	000	460-1400	460-1400	12-540
Total from EPA/NEI. all sectors reporting (metric tons)[3]	720 000	579 000		134 000	322 000	898 000
Shale extraction relative to total (%)	0.35-1.5	2.9-4.8		0.34-1.0	0.14-0.43	0.0013-0.060

[a]Combustion-based electric utilities and highway and off-highway vehicles generally constitute a large percentage of statewide emissions in EPA's 2008 NEI. For example, combustion-based electricity production, highway vehicles, and off-highway vehicles sectors statewide represent: 80% of NO_x (460 000 of 580 000 metric tons); 47% of $PM_{2.5}$ (63 000 of 130 000 metric tons); and 87% of SO_2 (780 000 of 900 000 metric tons). Combined, they are less significant for VOCs and PM_{10} (26% and 22% of statewide respectively).

Extraction activities, however, are not evenly distributed throughout the state, so it is instructive to look at the magnitude of emissions in the few counties where activities were concentrated in 2011. More than 20% of wells were found in one county and nearly 50% were in the top 3 counties; the 10 counties with the most development constituted nearly 90% of wells in the state (see table S.8 available at stacks.iop. org/ERL/8/014017/mmedia). The statewide extraction industry also produced VOC[12] and NOx [13] emissions equivalent to or larger than some of the largest single emitters in the state—GW-scale coal-based electric power plants. In the counties with the most activity, even the low-end of the NO_x emissions estimate ranges were 20–40 times higher than the level that would constitute a 'major' emissions source, although individually the new shale-related facilities are generally not subject to major source permit requirements. On the other hand, the magnitude of PM and SO2 emissions are much less significant relative to existing major sources, as the statewide totals imply[14].

[12]The top five and top twenty VOC emitters produce 252 metric tons per year and 542 tons per year, respectively, in 2008.

[13]For example, the range of estimates of emissions of NO_x is comparable to or larger than the emissions of the top four NO_x emitters in the state. These top four facilities reported emissions of about: 23 500; 22 200; 16 200; and 15 800 metric tons per year of NO_x. The facilities are 2.7, 1.7, 2.0, and 1.9 GW coal-fired power facilities, respectively.

[14]For example, the top four emitters of SO_2 in the state produce from 90 000 to 170 000 metric tons each, so even the high end of the estimates of SO_2 for the extraction industry are equivalent to less than a per cent of these.

[15]Calculation of the statewide damages of all major emitters involves estimating damages for each source individually, due to county-to-county variability of the damage function as well as accounting for each emissions source location and height, and is out of scope for this analysis.

Although the correlation with emissions is not direct, the total regional damages from the shale gas extraction industry are also expected to be small relative to statewide air pollution emissions damages15. For comparison, we estimate that the largest coal-fired power plant in Pennsylvania—while not the state's most polluting facility—alone produced about $75 million in damages in 2008. The four largest facilities—which included the top two SO2 emitters in the state—produced nearly $1.5 billion in damages in 2008. For the shale gas extraction industry, monetary damages were driven by significant levels of VOCs, NOx, and PM2:5, and the whole industry constituted less than 2%, 5%, and 1% for each of the pollutants, respectively, of total emissions in the state in 2008 from all industries reporting.

Because the relative damages will tend to be larger in the counties where shale gas extraction activities are concentrated, where population is relatively high, and where air quality is already a concern, it is also important to consider the county-level damage. For example, Washington County had the fifth largest number of wells (156) in 2011 but resulted in the highest damages, estimated at $1.2–8.3 million. Damage in this county represented about 20% of statewide damages from the extraction industry[16]. And while not typical of 2011 development, this example illustrates the potential impact of extraction when located in relatively populated areas[17].

[16]These damages were equivalent to about 11% of the damages from the largest electricity plant.

[17]In this case, Washington County is just south of Allegheny County and the city of Pittsburgh; previous development in the state occurred in more rural north and central Pennsylvania.

DISCUSSION

We estimate that total regional air-quality-related damages, at the level of development and production in Pennsylvania in 2011, ranged between $7.2 million and $32 million (table 6). However, extraction industry damages will not be constant over time or evenly distributed in space, and there are important policy implications of when and where emissions damages occur. Development emissions damages range from about $2.5 to $5.5 million, but the majority of annual attributable emissions will continue for the life of the well and associated compressor facilities. This is true despite the relatively high level of development activity in 2011 and the relatively low number of actively producing shale gas wells, compared to what is expected in coming years. At the low end of our estimates, 66% of total damages in 2011 were attributable to long-term activities; at the high end, more than 80% of damages occur in the years after the well is developed. Nor are most emissions associated with well-site activities. More than half of emissions damages from this industry come from compressor stations, which may serve dozens of individual wells, including conventional ones. Our estimates indicate that regulatory agencies and the shale gas industry, in developing regulations and best practices, should account for air emissions from ongoing, long-term activities and not just emissions associated with development, such as drilling and hydraulic fracturing, where much attention has been focused to date. Even if development slows in the Marcellus region, as it did in

2012, the long-term nature of these emission sources will mean that any new development will add to this baseline of emissions burden as more producing wells and compressor stations come online.

Additionally, most development activities do not constitute 'major sources' under federal air-quality regulations. Especially for those counties that already suffer from high levels of air pollution (i.e., those in or near Clean Air Act non-attainment status), these new activities may make meeting federal air-quality standards more difficult. This issue was raised in the context of the Haynesville Shale region, where authors noted that emissions could 'be sufficiently large that (they) . . . may affect the ozone attainment status' (Kemball-Cook et al 2010). It may be hard to limit these emissions through mechanisms such as permitting restrictions, which typically do not apply to mobile and minor stationary sources. Existing regulations may therefore not be well-suited for managing emissions from a substantial number of small-scale emitters. Proposals to aggregate industry sources should be carefully considered in terms of the appropriate unit of aggregation (e.g., by company, by geographic region) and any unintended consequences or perverse incentive they may create. One approach to reducing air emissions is to require the use of Best Available Technologies (BAT); for compressors, these include lean-burn engines, non-selective catalytic reduction, or electrification, measures often found to be cost-effective (Armendariz 2009). The various costs of meeting or exceeding BAT in Pennsylvania will likely be estimated to support updated compressor permit requirements in Pennsylvania in 2013.

It is worth stressing that a substantial portion of emissions estimated here are not specifically attributable to the 'unconventional' nature of shale gas. Natural gas compressor stations are necessary to produce and distribute natural gas from any source, from conventional to biomethane. So while the emissions levels estimated are non-trivial, they may not differ substantially from any other large-scale industrial emissions that impact regional air quality; it is the scale of the resource extraction or industrial activity that is likely to matter most. Additionally, the magnitude of the potential damages must be considered in the context of other external costs associated with this industry, as well as in terms of the potential benefits of shale gas use.

While statewide emissions from the extraction industry are relatively small compared to some other major sources of air pollution in the state (e.g., SO2 from GW-scale coal-fired power plants), these emissions sources are nevertheless a concern in regions of significant extraction activities. More detailed analyses, including regional data acquisition and consideration of site-specific variability, will be valuable in regions of intense extraction activity and for specific activities and pollutants shown in this analysis to be of most potential concern. And while significant uncertainty may exist for some potential risks of shale gas extraction, under current standard practices, shale gas extraction will be associated with non-trivial air pollution emissions.

ACKNOWLEDGMENTS

This research was funded by the RAND Corporation's Investment in People and Ideas program. Support for this program is provided, in part, by the generosity of RAND's donors and by the fees earned on client-funded research. We thank Pennsylvania Department of Environmental Protection (PA DEP) staff and one shale gas operator company for providing data and helpful discussions. Joe Osborne of the Group Against Smog and Pollution (GASP) provided access to compiled compressor station permits, and Nick Muller (Middlebury College) assisted with understanding and making use of the APEEP model. Thanks also to Henry Willis, Shanthi Nataraj, and Tom LaTourrette of RAND for helpful discussions. We also thank two anonymous reviewers for suggestions that significantly improved the manuscript.

REFERENCES

1.   Alvarez R A and Paranhos E 2012 Air pollution issues associated with natural gas and oil operations EM pp 22–5 (www.edf.org/sites/default/files/AWMA-EM-airPollutionFromOilAndGas. pdf)

2.   Armendariz A 2009 Emissions from Natural Gas Production in the Barnett Shale Area and Opportunities for Cost-Effective Improvements (Austin, TX: Environmental Defense Fund) (www.edf.org/sites/default/files/9235 Barnett Shale Report. pdf)

3.   Burnham A, Han J, Clark C E, Wang M, Dunn J B and Palou-Rivera I 2011 Life-cycle greenhouse gas emissions of shale gas, natural gas, coal, and petroleum Environ. Sci. Technol. 46 619–27

4. Chambers A K, Strosher M, Wootton T, Moncrieff J and McCready P 2008 Direct measurement of fugitive emissions of hydrocarbons from a refinery J. Air Waste Manag. Assoc. 58 1047–56

5. Considine T J, Watson R and Blumsack S 2011 The Pennsylvania Marcellus Natural Gas Industry: Status, Economic Impacts and Future Potential (University Park, PA: Pennsylvania State University, College of Earth and Mineral Sciences, Department of Energy and Mineral Engineering)

6. Fulton M, Mellquist N, Kitasei S and Bluestein J 2011 Comparing Life-Cycle Greenhouse Gas Emissions from Natural Gas and Coal (Frankfurt: Deutsche Bank and Worldwatch Institute)

7. Ground Water Protection Council and ALL Consulting 2009 Modern Shale Gas Development in the United States: A Primer (Oklahoma City, OK: US Department of Energy Office of Fossil Energy and National Energy Technology Laboratory)

8. Hultman N, Rebois D, Scholten M and Ramig C 2011 The greenhouse impact of unconventional gas for electricity generation Environ. Res. Lett. 6 044008

9. Jiang M, Griffin W M, Hendrickson C, Jaramillo P, VanBriesen J and Venkatesh A 2011 Life cycle greenhouse gas emissions of Marcellus shale gas Environ. Res. Lett. 6 034014

10. Kargbo D M, Wilhelm R G and Campbell D J 2010 Natural gas plays in the Marcellus shale: challenges and potential opportunities Environ. Sci. Technol. 44 5679–84

11. Kemball-Cook S, Bar-Ilan A, Grant J, Parker L, Jung J, Santamaria W, Mathews J and Yarwood G 2010 Ozone impacts of natural gas development in the haynesville shale Environ. Sci. Technol. 44 9357–63

12. Levi M A 2012 Comment on 'Hydrocarbon emissions characterization in the Colorado Front Range: a pilot study' by Gabrielle P´etron et al J. Geophys. Res. 117 D21203

13. Lu X, Salovaara J and McElroy M B 2012 Implications of the recent reductions in natural gas prices for emissions of CO2 from the US Power Sector Environ. Sci. Technol. 46 3014–21

14. Marcellus Shale Coalition 2012 American Natural Gas: A Source of Sustained Economic Growth (retrieved 11 October 2012 from http://marcelluscoalition.org/2012/08/american-natural-gas-asource- of-sustained-economic-growth/)

15.   McKenzie L M, Witter R Z, Newman L S and Adgate J L 2012 Human health risk assessment of air emissions from development of unconventional natural gas resources Sci. Total Environ. 424 79–87

16.   Mooney C 2011 The truth about fracking Sci. Am. 305 80–5 Muller N Z and Mendelsohn R 2007 Measuring the damages of air pollution in the United States J. Environ. Econom. Manag. 54 1–14

17.   Muller N Z and Mendelsohn R 2012 Efficient pollution regulation: getting the prices right: corrigendum (mortality rate update) Am. Econ. Rev. 102 613–6

18.   NRC (National Research Council) 2010 Hidden Costs of Energy: Unpriced Consequences of Energy Production and Use (Washington, DC: National Academies Press) (www.nap.edu/catalog.php?record id=12794)

19.   NYS DEC (New York State Department of Environmental Conservation) 2011 Revised Draft Supplemental Generic Environmental Impact Statement (EIS) on the Oil, Gas and Solution Mining Regulatory Program (www.dec.ny.gov/data/dmn/rdsgeisfull0911.pdf)

20.   PA DEP 2011 PA DEP Oil and Gas Reporting Website— Statewide Data Downloads by Reporting Period: Marcellus Only (retrieved July 2012 from www.paoilandgasreporting.state.pa.us/publicreports/Modules/DataExports/DataExports.aspx)

21.   PennEnvironment Research and Policy Center 2012 The Costs of Fracking: The Price Tag of Dirty Drilling's Environmental Damage (www.pennenvironment.org/sites/environment/files/   reports/The%20Costs%20of%20Fracking%20vPA 0.pdf)

22.   Pennsylvania Department of Environmental Protection 2011 Air Emissions Inventory for the Natural Gas Industry (retrieved 8 July 2012 from www.elibrary.dep.state.pa.us/dsweb/Get/ Document-86312/2700-FS-DEP4354.pdf)

23.   P´etron G, Frost G, Miller B R, Hirsch A I, Montzka S A, Karion A, Trainer M, Sweeney C, Andrews A E and Miller L 2012 Hydrocarbon emissions characterization in the Colorado Front Range: a pilot study J. Geophys. Res. 117 D04304

24.   Ramamurthy K 2012 Personal Communication on Air Emission Inventory Submissions to PA DEP

25. Shogren E 2011 Air Quality Concerns Threaten Natural Gas's Image (National Public Radio) (retrieved September 2012 from www.npr.org/2011/06/21/137197991/air-quality-concernsthreaten-natural-gas-image)

26. Skone T, Littlefield J, Eckard R, Cooney G and Marriott J 2012 Role of Alternative Energy Sources: Natural Gas Technology Assessment (NETL/DOE-2012/1539) (www.netl.doe.gov/ energy-analyses/refshelf/PubDetails.aspx?Action=View& PubId=435)

27. Steinzor N, Subra W and Sumi L 2012 Gas patch roulette: how shale gas development risks public health in Pennsylvania Earthworks Oil and Gas Accountability Project (www.earthworksaction.org/files/publications/Health-Report- Full-FINAL-sm.pdf)

28. US DOE Argonne National Labs (ANL) 2012 The Greenhouse Gases, Regulated Emissions, and Energy Use in Transportation (GREET) Model, GREET 2 2012 (http://greet.es.anl.gov/)

29. US DOE EIA 2012a Annual Energy Outlook 2012 (Washington, DC: US DOE)

30. US DOE EIA 2012b Monthly Natural Gas Gross Production Report (data for September 2012, www.eia.gov/oil gas/natural gas/ data publications/eia914/eia914.html, retrieved November 2012)

31. US EPA 1999 The Benefits and Costs of the Clean Air Act 1990 to 2010 (EPA Report to Congr. EPA-410-R-99-001) (Washington, DC: Office of Air and Radiation, Office of Policy, US Environmental Protection Agency) (www.epa.gov/oar/sect812/ 1990-2010/chap1130.pdf)

32. US EPA 2008 National Emissions Inventory (NEI) Database (from www.epa.gov/ttnchie1/net/2008inventory.html)

33. US EPA 2012 Air Rules for the Oil and Natural Gas Industry (retrieved July 2012 from www.epa.gov/airquality/oilandgas/actions.html)

34. US GAO 2012 Information on Shale Resources, Development, and Environmental and Public Health Risks (www.gao.gov/assets/ 650/647791.pdf)

35. Venkatesh A, Jaramillo P, Griffin W M and Matthews H S 2011 Uncertainty in life cycle greenhouse gas emissions from united states natural gas end-uses and its effects on policy Environ. Sci. Technol. 45 8182–9

36.   Weber C L and Clavin C 2012 Life cycle carbon footprint of shale gas: review of evidence and implications Environ. Sci. Technol. 46 5688–95

Chapter 7

The Direct use of Post-Processing Wood Dust in Gas Turbines

Alîne Doherty[1], Eilín Walsh[1],
and Kevin P. McDonnell[2]

[1]School of Biosystems Engineering, University College Dublin, Dublin, Ireland

[2]Animal and Crop Sciences, School of Agriculture, Food Science and Veterinary Medicine, University College Dublin, Dublin, Ireland

ABSTRACT

Woody biomass is a widely-used and favourable material for energy production due to its carbon neutral status. Energy is generally derived either through direct combustion or gasification. The Irish forestry sector is forecasted to expand significantly in coming years, and so the opportunity exists for the bioenergy sector to take advantage of the material for which there will be no demand from current markets. A by-product of wood processing, wood dust is the cheapest form of wood material available to the bioenergy sector. Currently wood dust is primarily processed into wood pellets for energy generation. Research was conducted on post-processing birch wood dust; the calorific value and the Wobbe In-dex were determined for a number of wood particle sizes and wood dust concentrations. The Wobbe Index determined for the upper explosive concentration (4000 g/m³) falls within range of that of hydrogen gas, and wood dust-air mix-tures of this concentration could therefore behave in a similar manner in a gas turbine. Due to its slightly lower HHV and higher particle density, however, alterations to the gas turbine would be necessary to accommodate wood dust to prevent abrasive damage to the turbine. As an unwanted by-product of wood processing the direct use of wood dust in a gas turbine for energy generation could therefore have economic and environmental benefits.

INTRODUCTION

Woody biomass is a widely-used material for energy production. Energy from wood is generally derived from direct combustion or gasification of wood in pellet or chip form. Due to its carbon neutral status, wood is a very desirable alternative to the use of fossil fuels as a source of primary energy. In addition to its environ-mental advantages, utilising woody material as a primary energy fuel can increase security of supply if the wood material is obtained from a domestic source.

The potential for wood to be used as a primary energy fuel in Ireland will be investigated, with particular em-phasis placed on the feasibility of using wood dust to generate heat and electrical energy. The current use of wood as an energy source will be outlined, the volumes of wood dust potentially available as a fuel source will be explored, and how this material can be used as a fuel, taking into consideration any amendments to existing equipment that may be necessary, will be examined.

In 2004 Sustainable Energy Ireland conducted a sur-vey of the bioenergy potential in Ireland and reported that the forestry sector had undergone significant expan-sion in previous years [1]. COFORD forecasted the Irish forestry sector to further expand to 5 million cubic me-tres by 2015 [2]: this expansion is evident in the almost 750,000 ha of forestry planted in 2010 [3]. It was con-sidered unlikely that the increased availability of wood material would be met by demand within current markets which means an alternative outlet must be found for the additional volume of woody material available. This ad-ditional material could be used to great benefit by the bioenergy industry to generate renewable energy from an indigenous fuel source. To take advantage of the in-creased availability and energetic potential of woody biomass from the expansion of the Irish forestry sector, a significant target for the integration of renewable electri-cal energy has been set: 30% co-firing in three peat- burning power stations by 2015 [4].

Wood-based bioenergy can take one of three forms: direct biomass, for example chipping of trees removed during thinning; indirect biomass, for example wood by-products such as sawdust and bark recovered from primary and secondary processing; or recovered waste wood, for example construction and demolition waste wood, old pallets, etc. Recovered wood generally has a lower moisture content than fresh biomass as a result of the original processing [1], which increases the effi-ciency of energy recovery from wood. Wood energy is currently used in Ireland for heat and electricity genera-tion: heat energy is primarily obtained from small-scale combustion in domestic boilers while electricity is gen-erated either by co-firing in power stations or by gasifi-cation or pyrolysis to generate a synthetic gas or liquid to be burned in a gas or steam turbine [1].

In Ireland energy is derived from wood material pri-marily by direct combustion in either pellet or chip form. Wood pellets are usually made from unprocessed, dry, waste wood which can be either hardwood or softwood in nature. Softwoods are more suitable than hardwoods for pellet production due to the higher content of lignin which acts as a binding agent; pellets made from hard-woods such as willow require the addition of a binding agent for durability [5]. Pellets have lower moisture con-tent and higher energy density than chips (17.0 GJ/t vs. 13.4 GJ/t, [6]) and produce a predictable fuel with mini-mum residual ash material [1], however processing costs are higher when producing wood pellets. Wood chips are usually produced from forestry logging residues and purpose-grown energy crops [6]. Wood chip production is more economical than pelleting as the required level of processing is lower and can be carried out on a small- scale, localised basis. Due to their lower energy density and higher moisture content [6], wood chips must un-dergo a degree of either active or passive drying prior to use in a boiler [6]. The higher moisture content must also be taken into consideration during storage to prevent degradation of the feedstock [7].

The cheapest form of woody material for energy gen-eration is wood dust produced as a by-product of proc-essing as there is essentially no further pre-treatment required. Wood dust is the most unfavoured by-product in current wood industries as it is difficult to handle and has the lowest energy density [2], so has the greatest po-tential as an available bioenergy feedstock for Ireland. Processing by-products including wood dust are cur-rently consumed on-site by a number of sawmills in Ire-land for the production of heat [1] to dry the incoming wood. In addition, the two largest pellet producers in Ireland convert approximately 200,000 tonnes of wood dust material into wood pellets for direct combustion for energy generation. Pellets can also be gasified or pyro-lysed rather than combusted; gasification is a more effi-cient process than combustion and therefore extracts more utilisable energy [8]. Wood dust itself can be gas-ified or pyrolysed to produce gas or liquid which can then be combusted to generate power. One potential problem with gasification of wood dust, however, is the transportation of ungasified particles into the turbine along with the produced gas [9]. Particulate matter mov-ing through the turbine can cause disruption of the tur-bine blades and can cause abrasion to the inner surfaces of the turbine [10].

Wood dust is comprised of cellulose, hemicellulose, lignin, and extractives. Lignin and extractives tend to be more prominent in softwoods than hardwoods, translat-ing into a higher heating value in softwoods [11]. Wood dust has a relatively high volatiles content (60% - 70%) and low heating value (17 - 18 MJ/kg) and thus does not combust efficiently in conventional combustors [12]. From an ignition point of view, the ignition temperature of wood dust is lower than for whole wood: between 204°C and 260°C [13] and between 350°C and 600°C [14], respectively. Ignition point is influenced by multi-ple factors including wood source, moisture content, par-ticle size, molecular composition, dispersion, and con-centration [13]. The smaller particle size associated with dust is therefore advantageous for ignition, hence the lower ignition point.

Moisture content is one of the most important factors when considering wood dust as a fuel, both with regard to the potential extractable energy and with regard to the ignition point. The higher the moisture content of a feed-stock for energy generation, the greater the required ini-tial energy input to evaporate the moisture [15]. Fresh wood dust has much higher moisture content than indus-try-derived wood dust due to drying during preparation for industrial use. The moisture content of fresh wood material can be as high as 60% compared to approxi-mately 10% for industry-dried wood dust [11]. Wood dust moisture content influences five main characteristics of wood dust explosions: maximum explosion pressure (P_{max}); maximum rate of pressure rise (K_{St}); minimum ignition energy (MIE); minimum explosible dust concen-tration (MEC); and minimum ignition temperature (MIT). Increasing moisture content and particle size decrease the maximum explosion pressure and maximum rate of pres- sure rise and increase the minimum ignition energy [15]. In addition, increasing moisture content and particle size increase minimum explosible dust concentration and minimum ignition temperature.

The forecasted expansion of the forestry sector has positive implications for biomass-based bioenergy in Ireland. Due to the high volumes of wood dust produced as a result of wood processing and therefore available without additional treatment such as the necessary proc-essing to produce wood pellets, it is considered that wood dust is an under-exploited source of bioenergy ma-terial in Ireland. The aim of this research, therefore, is to investigate the potential use of wood dust directly in gas turbines, *i.e.* the use of wood dust to generate energy without first producing wood pellets.

METHODS AND MATERIALS

To determine the advantages of direct wood dust com-bustion in a gas turbine system, three dust particle sizes were investigated: 425 μm, 150 μm, and 63 μm. To iso-late dust particles of these specific dimensions a mixed sample of Russian and Irish (approximately 90% and 10%, respectively) birch plywood derived from the fur-niture industry was sieved and the relevant fractions were extracted. Moisture content of the dust samples (n = 3) was measured by drying in a convection oven to a con-stant weight.

The higher heating value (HHV) of the wood dust was determined using a Parr 6400 (Parr Instrument Company, Moline, Illinois, USA) bomb calorimeter. The HHV by volume (HHV$_v$) was used in conjunction with the spe-cific gravity of each wood dust-air concentration to determine the Wobbe Index of each wood dust-air concen-tration using the following equations:

$$HHV_v = HHV \times \rho_{bulk}$$

(1)

% Concentration by volume

$$(C_v) = \left(\frac{Concentration}{\rho_{bulk}} \right)$$
$$\times 100$$

(2)

Specific gravity of wood dust-air concentration (G$_s$)

$$= \frac{C_v \times \rho_{bulk} + (1 - C_v) \times \rho_{air}}{\rho_{air}}$$

(3)

Where $_{bulk}$ = bulk density of wood sample, and $_{air}$ = density of air (1.2041 kg/m³).

Wobbe Index was therefore calculated for each wood dust-air concentration as:

$$Wobbe\ Index = \frac{HHV_V \times C_V}{\sqrt{G_S}}$$

(4)

Simultaneous thermal analysis (STA) was conducted to determine ignition points, weight loss due to ignition, and loss of volatiles at each particle size and wood-dust air concentration. This analysis was conducted using a Rheometric Scientific STA 1000 (Rheometric Scientific Inc, Piscataway, New Jersey, USA) apparatus on wood dust samples of each particle size at a number of wood dust-air concentrations: 50 g/m³, 500 g/m³, and 4000 g/m³. This gave a total of nine samples, each analysed in triplicate (Table 1).

Table 1: Experimental wood dust-air concentrations and particle sizes tested in STA

	Minimum explosive concentration 50 g/m³	Stoichiometric concentration 500 g/m³	Upper explosive concentration 4000 g/m³
Maximum particle size	425μm	425μm	425μm
Mid particle size	150μm	150μm	150μm
Minimum particle size	63μm	63μm	63μm

RESULTS AND DISCUSSION

The moisture content of the wood dust was determined to be 4%, a value which was expected due to the post-con- sumer nature of the wood dust. As was described earlier, low moisture content corresponds to a higher heating value, a lower ignition temperature, and a greater loss of volatiles [1]. A low moisture content such as that ob-served here increases the likelihood of a wood dust ex-plosion occurring and enhances the kinetics of the reac tion. The bulk density for the wood dust was also deter-mined during this research and was calculated to be 380.23 kg/m³.

Results from the simultaneous thermal analysis are shown in Table 2 and indicate that wood dust-air con-centration and particle size have a considerable influence on points of ignition and weight loss due to ignition. It was observed that ignition temperature increased with increasing particle size and that weight loss due to igni-tion was greatest at 150 μm particle size. The analysis indicates that at all three concentrations examined the smallest particle size (63 μm) required the lowest ignition energy and therefore recorded the lowest point of igni-tion.

For all particle sizes the lowest point of ignition was recorded for the stoichiometric concentration of 500 g/m³. Minimum explosive concentration and upper explosive concentration had the greatest percent weight loss at 425 μm particle size whereas percent weight loss was greatest at 150 μm particle size at stoichiometric concentrations. For all concentrations the greatest weight loss was re-corded for 63 μm particle size, and for all particle sizes the most pronounced weight loss was recorded at 50 g/m3 concentration. A more consistent pattern of weight loss was observed for the stoichiometric and upper explosive concentrations than that observed for the minimum ex-plosive concentration, which indicates more stable com-bustion at higher wood dust-air concentrations.

Fungtammasan *et al.* [12] reported the higher heating value by mass (HHV_m) of wood dust to be approximately 17 - 18 MJ/kg. Due to the low moisture content and the species used in this research, a HHV_m of 19.16 MJ/kg was recorded. Present day turbines can typically operate using gases with a HHV_m between 9.4 MJ/kg (CO) and 54 MJ/kg (natural gas) [16], therefore a HHV_m of 19.18 MJ/kg is well within operational range. The Wobbe In-dex can be used to determine the interchangeability of wood dust-air mixtures with other operational gases. The results obtained indicate that both the stoichiometric and upper explosive concentrations of wood are within the limits of Wobbe Indices of current practical gaseous fu-els (Table 2). The recorded Wobbe Index for the upper explosive concentration falls within range of the HHV of hydrogen gas, and wood dust-air mixtures of this con-centration could therefore behave in a similar manner in a gas turbine. Due to its slightly lower HHV_m and higher particle density, however, alterations to the gas turbine would be necessary to accommodate wood dust as an energy fuel.

The primary adjustment necessary would be a size al-teration: fuels with lower heating values require a greater volume of fuel to meet temperatures achieved by fuels with higher heating values, and thus require a longer combustion zone within the turbine [16]. In addition, to prevent damage resulting from the use of a more abrasive fuel, vertically-mounted cyclone combustors could be used to burn fuels with a range of heating values which ensure adequate particle entrapment. Conical-shaped com- bustors collect particles at the base and avoid particle infiltration to the turbine blades [9]. It is further recom-mended that the combustor and blades be lined to protect the blades against abrasion.

Table 2: Results of calorific value and simultaneous thermal analysis conducted for each wood dust-air concentration and particle size investigated

Wood dust-air concentration	Minimum explosive concentration 50 g/m³	Stoichiometric concentration 500 g/m³	Upper explosive concentn 4000 g/m³
HHV$_y$ (MJ/m³)	7293	7293	7293
Specific gravity	1.041	1.414	4.309
Wobbe Index (MJ/m³)	0.940	8.066	36.960
Ignition point (°C)	249.85	252.14	240.04
Weight loss due to ignition (%)	7.761	0.625	0.175
Particle size (µm)	63	150	425
Ignition point (°C)	235.722	246.847	259.469
Weight loss due to ignition (%)	2.343	3.211	3.006

CONCLUSIONS

The results of this study show the use of wood dust as a primary fuel in gas turbines for power generation to be both feasible and advantageous. At the upper explosive concentration investigated, the Wobbe Index was found to be similar to that of hydrogen gas and, with small enough particle size (≤63 µm), wood dust could behave similarly in a gas turbine. Gas turbine design alterations would be required to ensure proper injection, dispersion, and mixing of the wood dust in the turbine as well as to protect the turbine from abrasion caused by the wood dust particles.

The moisture content of the wood dust used in this in-vestigation was found to be 4%, which was unsurprising given the post-processing nature of the wood dust. In Ireland there are vast quantities of wood dust produced annually during processing of wood into consumer products which would have a similar moisture content. This abundance of waste woody material coupled with ad- vancing bioenergy technology could contribute to reliev- ing the import dependency faced by Ireland for primary energy by generating energy from a domestic source which is also carbon neutral.

REFERENCES

1.   SEI, "Bioenergy in Ireland," Sustainable Energy Ireland, Dublin, 2004.

2.   Programme of Competitive Forestry Research for Devel- opment, "Strategic Study: Maximising the Potential of Wood Use for Energy Generation in Ireland," 2004. http://www.seai.ie/Renewables/ Bioenergy/Maximising_the_potential_of_wood_energy,_Coford. pdf

3.   Teagasc, "Forestry Statistics 2010," Teagasc, Carlow, 2010.

4.   DCMNR, "Bioenergy Action Plan for Ireland," Depart-ment of Communications, Energy and Natural Resources, Dublin, 2007.

5.   M. Peksa-Blanchard, P. Dolzan, A. Grassi, J. Heinimö, M. Junginger, T. Ranta and A. Walter, "IEA Bioenergy Task 40: Global Wood Pellets Markets and Industry: Policy Drivers, Market Status and Raw Material Potential," In-ternational Energy Agency, Paris, 2007.

6.   Sustainable Energy Authority of Ireland, "Wood Fuel and Supply Chain," 2011. http://www.seai.ie/Renewables/Bioenergy/ Sources/Wood_Energy_and_Supply_Chain/Fuel_and_Supply_ Chain/

7.   M. R. Wu, D. L. Schott and G. Lodewijks, "Physical Properties of Solid Biomass," Biomass and Bioenergy, Vol. 35, No. 5, 2011, pp. 2093-2105. doi:10.1016/j.biombioe.2011.02.020

8.   GTC, "Gasification: The Waste-to-Energy Solution," Gasifi- cation Technologies Council, Arlington, 2012.

9.   C. Syred, A. Griffiths and N. Syred, "Gas Turbine Com-bustor with Integrated Ash Removal for Fine Particu-lates," Proceedings ASME Turbo Expo, Vienna, 14-17 June 2004, pp. 1-9.

10. D. J. Flynn, J. J. Dillon, P. B. Desch and T. S. Lai, "The NALCO Guide to Boiler Failure Analysis," 2nd Edition, McGraw Hill, Inc., New York, 2011.

11. K. W. Ragland, D. J. Aerts and A. J. Baker, "Properties of Wood for Combustion Analysis," Bioresource Technol-ogy, Vol. 37, 1991, pp. 161-168.doi:10.1016/0960 8524(91)90205-X

12. B. Fungtammasan, P. Jittreepit, J. Torero and P. Joulain, "An Experimental Study of the Combustion Characteris-tics of Sawdust in a Cyclone Combustor," Proceedings European— ASEAN Conference on Combustion of Solids and Treatment of Products, Hua Hin, 16-17 February 1995, pp. 1-18.

13. Weyerhaeuser Company, "Wood and Wood Dust (With-out Chemical Treatments or Resins/Adhesives). Material Safety Data Sheet," Weyerhaeuser Company, Washington, 2010.

14. V. Babrauskas, "Ignition of Wood: A Review of the State of the Art," Proceedings Interflam, 9th International Fire Science and Engineering Conference, Edinburgh, 17-19 September 2001, pp. 71-88.

15. R. K. Eckhoff, "Dust Explosions in the Process Industry," Gulf Professional Publishing, Massachusetts, 2003.

16. M. P. Boyce, "Gas Turbine Engineering Handbook," Vol. 4, Butterworth-Heinemann, Woburn, 2011.

Chapter 8

Characterization of Trace Gases Measured Over Alberta Oil Sands Mining Operations: 76 Speciated C_2–C_{10} Volatile Organic Compounds (VOCs), CO_2, CH_4, CO, NO, NO_2, No_y, O_3 and SO_2

I. J. Simpson[1], N. J. Blake[1], B. Barletta[1], G. S. Diskin[2],
H. E. Fuelberg[3], K. Gorham[1], L. G. Huey[4], S. Mein-
ardi[1], F. S. Rowland[1], S. A. Vay[2], A. J. Weinheimer[5], M.
Yang[1,2], and D. R. Blake[1]

[1]Dept. of Chemistry, University of California-Irvine, Irvine, CA 92697, USA

[2]NASA Langley Research Center, Hampton, VA 23681, USA

[3]Dept. of Meteorology, Florida State University, Tallahassee, FL 32306, USA

[4]School of Earth & Atmospheric Sciences, Georgia Institute of Technology, Atlanta, GA 30332, USA

[5]National Center for Atmospheric Research, 1850 Table Mesa Dr., Boulder, CO 80305, USA

ABSTRACT

Oil sands comprise 30% of the world's oil reserves and the crude oil reserves in Canada's oil sands deposits are second only to Saudi Arabia. The extraction and processing of oil sands is much more challenging than for light sweet crude oils because of the high viscosity of the bitumen contained within the oil sands and because the bitumen is mixed with sand and contains chemical impurities such as sulphur. Despite these challenges, the importance of oil sands is increasing in the energy market. To our best knowledge this is the first peer-reviewed study to characterize volatile organic compounds (VOCs) emitted from Alberta's oil sands mining sites. We present high-precision gas chromatography measurements of 76 speciated C_2–C_{10} VOCs (alkanes, alkenes, alkynes, cycloalkanes, aromatics, monoterpenes, oxygenated hydrocarbons, halocarbons and sulphur compounds) in 17 boundary layer air samples collected over surface mining operations in northeast Alberta on 10 July 2008, using the NASA DC-8 airborne laboratory as a research platform. In addition to the VOCs, we present simultaneous measurements of CO_2, CH_4, CO, NO, NO_2, NO_y, O_3 and SO_2, which were measured in situ aboard the DC-8.

Carbon dioxide, CH_4, CO, NO, NO_2, NO_y, SO_2 and 53 VOCs (e.g., non-methane hydrocarbons, halocarbons, sulphur species) showed clear statistical enhancements (1.1– 397x) over the oil sands compared to local background val-ues and, with the exception of CO, were greater over the oil sands than at any other time during the flight. Twenty halocarbons (e.g., CFCs, HFCs, halons, brominated species) either were not enhanced or were minimally enhanced (<10%) over the oil sands. Ozone levels remained low because of titration by NO, and three VOCs (propyne, furan, MTBE) remained below their 3 pptv detection limit throughout the flight. Based on their correlations with one another, the compounds emitted by the oil sands industry fell into two groups: (1) evaporative emissions from the oil sands and its products and/or from the diluent used to lower the viscosity of the extracted bitumen (i.e., C_4–C_9 alkanes, C_5–C_6 cycloalkanes, C_6–C_8 aromatics), together with CO; and (2) emissions associated with the mining effort, such as upgraders (i.e., CO_2, CO, CH_4, NO, NO_2, NO_y, SO_2, C_2–C_4 alkanes, C_2– C_4 alkenes, C9 aromatics, short-lived solvents such as C_2C_{l4} and C_2HCl_3, and longer-lived species such as HCFC-22 and HCFC-142b). Prominent in the second group, SO_2 and NO were remarkably enhanced over the oil sands, with maximum mixing ratios of 38.7 ppbv and 5.0 ppbv, or 383× and 319× the local background, respectively. These SO_2 levels are comparable to maximum values measured in heavily polluted megacities such as Mexico City and are attributed to coke combustion. By contrast, relatively poor correlations between CH_4, ethane and propane suggest low levels of natural gas leakage despite its heavy use at the surface mining sites. Instead the elevated CH_4 levels are attributed to methanogenic tailings pond emissions.

In addition to the emission of many trace gases, the natural drawdown of OCS by vegetation was absent above the surface mining operations, presumably because of the widespread land disturbance. Unexpectedly, the mixing ratios of _-pinene and _-pinene were much greater over the oil sands (up to 217 pptv and 610 pptv, respectively) than over vegetation in the background boundary layer (20±7 pptv and 84±24 pptv, respectively), and the pinenes correlated well with several industrial tracers that were elevated in the oil sands plumes. Because so few independent measurements from the oil sands mining industry exist, this study provides an important initial characterization of trace gas emissions from oil sands surface mining operations.

INTRODUCTION

Hydrocarbons are the basis of oil, natural gas and coal. Crude oil contains hydrocarbons with five or more carbon atoms (i.e., $\geq$_C_5), with an average composition of alkanes (30%), cycloalkanes (49%), aromatics (15%) and asphaltics (6%) (Alboudwarej et al., 2006). Oil sands comprise 30% of total world oil reserves (Alboudwarej et al., 2006) and are a mixture of sand, water, clay and crude bitumen – a thick, sticky extra-heavy crude oil that is "unconventional", i.e., does not flow and cannot be pumped without heating or dilution. With 179 billion barrels in its deposits, most of which occurs as oil sands, Canada has the world's 2nd largest crude oil reserves after Saudi Arabia and is currently the world's 7th largest producer of crude oil, generating about 1 million barrels day−1 with almost 4 million barrels day−1 expected by 2020 (www.canadasoilsands.ca). Alberta is Canada's largest oil producer and has oil sands deposits in three relatively remote regions: Athabasca (which is serviced by Fort McMurray), Peace River and Cold Lake (Fig. 1). Shallow oil sands deposits (<75m deep) in the Athabasca region can be surface mined, which has disturbed about 600 km² of land and comprises _20% of Alberta's oil sands production (www.oilsands.alberta.ca). Each barrel of oil generated from surface mining requires about 2 tons of oil sands, 3 barrels of water – much of which is re-used – and 20m3 of natural gas (http://www.neb. gc.ca/ clf-nsi/rnrgynfmtn/nrgyrprt/lsnd/pprtntsndchllngs20152006/ qapprtntsndchllngs20152006-eng.html). Deeper deposits occur in all three regions and require other in situ recovery methods such as steam injection and have higher natural gas requirements, but they have the advantage of much less land disturbance and no need for tailings ponds (www.oilsands.alberta.ca).

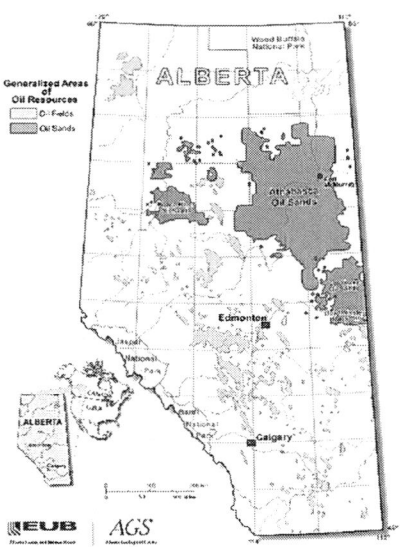

Figure 1: Alberta's oil sands deposits in the Athabasca, Peace River and Cold Lake regions (brown areas) (http://www.energy.alberta.ca/ Oil/pdfs/oil re-sources Map.pdf).

Oil sands processing extracts the bitumen from the oil sands and upgrades it into synthetic crude oil. The surface mining operators (e.g., Syncrude Canada, Suncor Energy, Albian Sands Energy) extract the bitumen using a hot water process (e.g., www.syncrude.ca/users/folder.asp?

FolderID=5918). Solvents/diluents are used to remove water and solids from the extracted bitumen and also decrease the bitumen's viscosity so that it can be piped to upgraders and refineries for processing. Diluent such as naphtha is used, which is a flammable liquid mixture of C_3–C_{14} hydrocarbons with major fractions of n-alkanes (e.g., heptanes, octane, nonane) and aromatics (e.g., benzene, toluene, ethylbenzene, xylenes) (Siddique et al., 2007). Paraffinic solvents/diluents may also be used, which consist of pentanes and hexanes (Siddique et al., 2006; D. Spink, personal communication, 2010). Upgraders crack and separate the bitumen into a number of fractions, e.g., fuel gas, synthetic crude oil, carbon (coke), and sulphur. There are upgraders at some of the surface mining sites north of Fort McMurray, at an in situ site south of Fort McMurray, and at downstream industrial centers such as the industrial heartland in Fort Saskatchewan, Alberta. Every $1m^3$ of oil sands also generates $4m^3$ of tailings waste, which includes both residual bitumen and diluent (Holowenko et al., 2000; Siddique, 2006). For example the Syncrude Mildred Lake tailings pond contains about 200 million m^3 of mature fine tailings and became methanogenic in the 1990s (Holowenko et al., 2000; Siddique et al., 2008). The methanogens are believed to anaerobically degrade certain components of naphtha (C_6– C_{10} n-alkanes and aromatics such as toluene and xylenes) into methane (CH_4), at rates of up to 10 gCH_4 m^{-2} d^{-1} or 40 million LCH_4 d^{-1} (Holowenko et al., 2000; Siddique et al., 2006, 2007, 2008; Penner and Foght, 2010).

Despite the emerging importance of oil sands in the energy market, characterizations of the emissions of volatile organic compounds (VOCs) and other trace gases from the oil sands industry are extremely scarce in the peer-reviewed literature. Whereas downstream oil sands upgrading and refining facilities can be monitored by ground-based studies (e.g., Mintz and McWhinney, 2008), independent studies of oil sands mining emissions are particularly difficult because the mining operations are not accessible to the public (Timoney and Lee, 2009). Instead, the majority of oil sands studies are reported in the so-called "grey literature", consisting of discipline-specific reports, industrial monitoring reports, reports by industrially-controlled consortia, and reports commissioned by non-governmental organizations (Timoney and Lee, 2009). Air quality in the oil sands surface mining air shed is monitored locally by the Wood Buffalo Environmental Association (WBEA), which is a multi-stakeholder organization that represents industry, environmental groups, government, communities, and Aboriginal stakeholders. Longterm monitoring data and reports are available at the WBEA website (www.wbea.org) but these data have not been published in the peer-reviewed literature.

We present independent observations of 76 speciated C_2– C_{10} VOCs, CH_4, sulphur dioxide (SO_2), carbon dioxide (CO_2), carbon monoxide (CO), ozone (O_3), nitric oxide (NO), nitrogen dioxide (NO_2) and total reactive nitrogen (NO_y =$NO + NO_2$ + HNO_3 + PANs + other organic nitrates + HO_2NO_2 + $HONO$ + NO_3 + $2 \times N_2O_5$ + particulate NO^-_3 + . . .) near surface oil sands mining and upgrading operations in Alberta's Athabasca region. Although present in trace quantities in the Earth's atmosphere, these gases drive the atmosphere's chemistry and radiative balance (Forster et al., 2007 and references therein). For example hydrocarbons and nitrogen oxides (NO_x =$NO+NO_2$) are key atmospheric constituents that can react together in the presence of sunlight to form tropospheric O_3, itself a greenhouse gas and air pollutant. Many hydrocarbons (e.g., benzene, toluene) can also be toxic or carcinogenic. Sulphur dioxide is produced by industrial processes including petroleum combustion and can contribute to photochemical smog and acid rain. Our measurements were made on 10 July 2008 during a 17 min boundary layer (BL) flight leg over the Alberta oil sands using the NASA DC-8 aircraft as a research platform (http:// airbornescience.nasa.gov/platforms/aircraft/dc-8.html). The VOCs were collected in seventeen 45 s integrated whole air samples that were subsequently analyzed at the University of California, Irvine (UC-Irvine). The remaining trace gases were sampled continuously, with CO_2, CH_4, CO, NO, NO_2, NO_y and O_3 reported every 1 s and SO_2 reported every 30 s. Although clearly limited in temporal and spatial extent, to our best knowledge these measurements represent the first independent characterization of speciated VOCs and many other trace gases from oil sands mining in the peer-reviewed literature.

EXPERIMENTAL

Boundary layer air was sampled over the Athabasca surface mines as part of the 2008 Arctic Research of the Composition of the Troposphere from Aircraft and Satellites (ARCTAS) field mission (www-air.larc.nasa.gov/missions/arctas/arctas. html). The summer deployment of ARCTAS was based in Cold Lake, Alberta (54°25' N; 110°12 W) and included eight 8-h science flights (Flights 17–24) from 29 June–13 July 2008. Although the major focus of the summer phase of ARCTAS was boreal biomass burning emissions (Jacob et al., 2009), on 10 July 2008 (Flight 23) there was an opportunity for the DC-8 to make two descents into the BL over northeast Alberta as part of a return transit flight from Thule, Greenland to Cold Lake (Fig. 2). During the first BL excursion (Leg 7) the DC-8 circled over the Athabasca oil sands mining area between 11:27–11:44 local time, at altitudes between 720–850m and within an area bounded by 56_340–57_090 N and 111°01–111°50W. The aircraft flew over both boreal forest and cleared industrial land including tailings ponds, tailings sand and upgrader facilities (Fig. 2b). Ten-day backward trajectories show that during Leg 7 the air masses arrived at the aircraft's pressure level from the west (Fig. 3a). The second BL run was flown shortly after, from 12:00–12:15 local time at altitudes between 980– 1410m (Leg 9). Leg 9 was an intercomparison flight leg between the DC-8 and the NASA P-3B aircraft that occurred in generally clean air approximately 1° further south (55_390–56_160 N and 112°44–113°47 W), i.e., not over the oil sands. The ten-day backward trajectories for Leg 9 show that the sampled air masses also arrived from the west and not from the oil sands mines to the north (Fig. 3b). Therefore, even though Leg 9 occurred at higher altitudes within the BL, we believe it provides a reasonable concurrent local background against which to compare the oil sands VOC enhancements (see additional discussion in Sect. 3.2.3).

Volatile Organic Compounds (VOCs)

Airborne Whole Air Sampling

UC-Irvine has measured speciated VOCs from diverse environments for more than 30 years, using both ground-based and airborne platforms (e.g., Blake and Rowland, 1988, 1995; Colman et al., 2001; Blake et al., 2003, 2008; Katzenstein et al., 2003; Barletta et al., 2005, 2009; Simpson et al., 2002, 2006). Our sampling technique collects whole air samples (WAS) into 2-L electropolished, conditioned stainless steel canisters each equipped with a Swagelok Nupro metal bellows valve (Solon, OH). The electropolishing minimizes the surface area and any surface abnormalities, and the canister conditioning (baking the cans in humidified air at ambient pressure and 225 °C for 12 h) forms an oxidative layer on the interior surface that further passivates the canister walls. The DC-8 payload included 168 of our 2-L air sampling canisters for each flight of the ARCTAS mission.

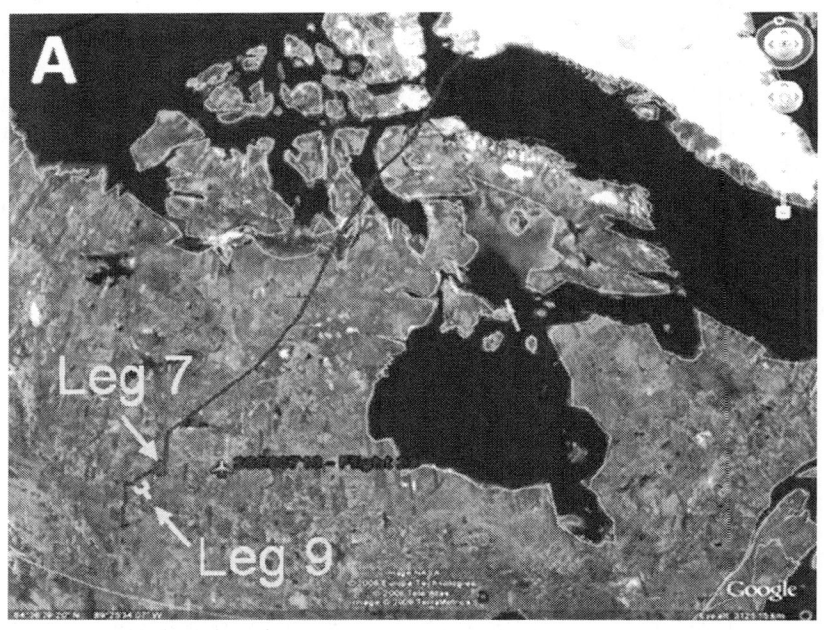

(a)

(b)

Figure 2: (a) Flight track of the NASA DC-8 aircraft during Flight 23, a transit flight from Thule, Greenland to Cold Lake, Alberta on 10 July 2008. The DC-8 circled within the boundary layer over the oil sands for 17 min (Leg 7) then descended back into background boundary layer air 16 min later (Leg 9). (b) Detail of the flight path during Leg 7. The locations of selected samples (in which maximum values were measured) are highlighted in yellow. The prevailing wind direction was from the southwest quadrant, and samples 4, 5 and 6 were collected directly downwind of the oil sands operations (the grey patch just above the center of the figure).

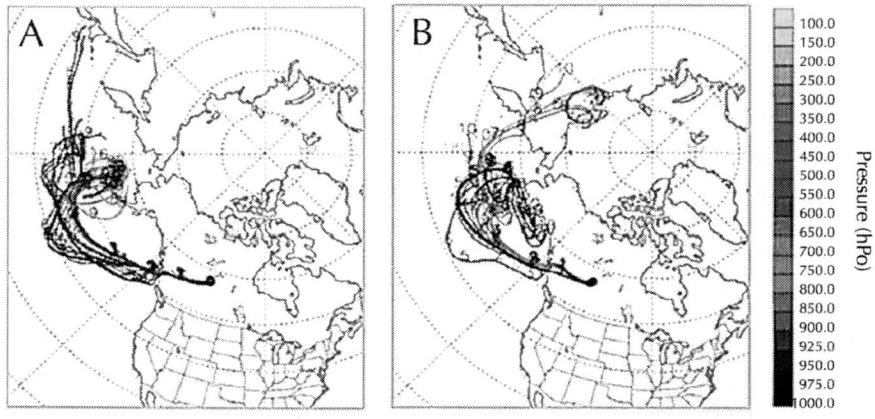

Figure 3: Ten-day backward kinematic trajectory plots starting at flight level pressure for the two boundary layer excursions flown on 10 July 2008 (a) over the Alberta oil sands from 11:27–11:44 local time (Leg 7), and (b) in background boundary layer air from 12:00–12:14 local time (Leg 9). Computational details are given in Fuelberg et al. (2010).

To prepare the canisters for field use, they are taken to the University of California Crooked Creek Station in the Sierra Nevada mountains (altitude=10 200 ft) for a pumpand- flush procedure that is repeated ten times, in which each canister is pressurized to 40 psig with ambient air and then vented to ambient pressure. Next the canisters are returned to our UC-Irvine laboratory where they are evacuated to 10^{-2} Torr and then pressurized to 1000 Torr with ultra-high purity helium before a final evacuation to 10^{-2} Torr (E2M12 dual-stage vacuum pumps, Edwards Vacuum, Wilmington, MA). Lastly each canister is humidified by adding ~17 Torr of purified water (the approximate vapour pressure of water at room temperature) to minimize surface adsorption and improve the reproducibility of our analytical split ratios during laboratory analysis (see Sect. 2.1.2). Rigorous sensitivity tests have shown that alkene growth in our passivated canisters is limited to a maximum of 0.2 pptv day^{-1} and all other compounds reported here are stable over the typically short period that the cans are stored before analysis (Sive, 1998). By analyzing the Alberta samples within 7 days of collection, we were able to limit the size of any alkene artefact to 1.4 pptv or less (i.e., to negligible values).

During each flight the whole air sampling is manually controlled and uses a stainless steel dual head metal bellows pump (MB-602, Senior Aerospace Metal Bellows, Sharon, MA) that is configured in series to draw outside air into a window-mounted 1/400 forward-facing inlet, through our air sampling manifold, and into one of the 168 canisters until it is filled to 40 psig. During Leg 7 we collected seventeen 45- second air samples, or approximately one sample per minute. During Leg 9 we collected 1-min integrated air samples every 3 min, for a total of 6 samples.

Laboratory Analysis of VOCs

After Flight 23 the whole air samples were returned to our UC-Irvine laboratory for analysis using three gas chromatography (GC) ovens coupled with a suite of detectors that together are sensitive to the 76 C_2–C_{10} VOCs that we seek to measure. We use two flame ionization detectors (FIDs) to measure hydrocarbons, two electron capture detectors (ECDs) for halocarbons, and a quadrupole mass spectrometer detector (MSD) for sulphur compounds.

Complete analytical details are given in Colman et al. (2001). For each sample a 1520 cm^3 sample aliquot is introduced into the analytical system's manifold and passed over glass beads contained in a loop maintained at liquid nitrogen temperature. A mass flow controller (Brooks Instrument; Hatfield, PA; model 5850E) keeps the flow below 500 cm^3 min^{-1} to ensure complete trapping of the less volatile sample components (e.g., VOCs) while allowing more volatile components (e.g., N_2, O_2, Ar) to be pumped away. The less volatile species are re-volatilized by immersing the loop in hot water (80 _C) and are then flushed into a helium carrier flow. The sample flow is split into five streams, with each stream chromatographically separated on an individual column and sensed by a single detector, namely: (1) a DB-1 fused silica capillary column (J&W Scientific; 60 m, I.D. 0.32 mm, film 1 mm) connected to an FID; (2) a GS-Alumina PLOT column (J&W Scientific; 30 m, I.D. 0.53 mm) spliced with a DB-1 fused silica capillary column (J&W Scientific; 5 m, I.D. 0.53 mm, film 1 mm) and connected to an FID; (3) an Rtx-1701 fused silica capillary column (Restek; 60 m, I.D. 0.25 mm, film 0.50 mm) connected to an ECD; (4) a DB-5 column (J&W Scientific; 30 m, I.D. 0.25 mm, film 1 mm) spliced with an Rtx-1701 column (Restek; 5 m, I.D. 0.25 mm, film 0.5

mm) and connected to an ECD; and (5) a DB-5ms fused silica capillary column (J&W Scientific; 60 m, I.D. 0.25 mm, film 0.5 mm) connected to an MSD. The split ratios are highly reproducible as long as the specific humidity of the injected air is above 2 g-H_2O/kg-air, which we ensure by adding purified water into each canister (Sect. 2.1.1). The signal from each FID and ECD is output to a personal computer and digitally recorded using Chromeleon Software; the MSD output signal uses Chemstation software. To ensure that the measurements are of the highest calibre, each peak of interest on every chromatogram is individually inspected and its integration is manually modified. For Flight 23 more than 10 000 peaks were hand-modified.

Calibration is an ongoing process whereby new standards are referenced to older certified standards, with appropriate checks for stability and with regular inter-laboratory comparisons. The hydrocarbons standards are NIST-traceable and the halocarbon standards are either NIST-traceable or were made in-house and have been compared to standards from other groups such as NOAA/ESRL. Multiple standards are used during sample analysis, including working standards (analyzed every four samples) and absolute standards (analyzed twice daily). Here we used working standards collected in the Sierra Nevada mountain range. International intercomparison experiments have demonstrated that our analytical procedures consistently yield accurate identification of a wide range of blindly selected hydrocarbons and produce excellent quantitative results (e.g., Apel et al., 1999, 2003). Table 1 shows the limit of detection (LOD) and the measurement precision and accuracy for each VOC. Even though our LOD is conservative, the accuracy and precision of many species does deteriorate as we approach this limit.

SO_2, CO_2, CH_4, CO, NO, NO_2, NO_y and O_3

In addition to VOCs measured by UC-Irvine, we also present SO_2, CO_2, CO, CH_4, NO, NO_2, NO_y and O_3 mixing ratios, which were measured in situ aboard the DC-8 by four research teams each using fast-response, high precision, continuous real-time instruments. The measurement precision and accuracy of these compounds are given in Table 1. The time response is 1 s for all compounds except NO and

NO_2, which have a 3 s time response (SO_2 has a 1 s time response but was reported as a 30 s average). Briefly, SO_2 was measured using the Georgia Tech Chemical Ionization Mass Spectrometer (GT-CIMS) instrument, which uses SF^-_6 ion chemistry to selectively ionize SO_2 (Kim et al., 2007). Carbon dioxide CO_2 was measured using the NASA Langley Atmospheric Vertical Observations of CO_2 in the Earth's Troposphere (AVOCET) instrument, which uses a modified Li- Cor model 6252 differential, non-dispersive infrared (NDIR) gas analyzer at the 4.26 µm CO_2 absorption band (Vay et al., 1999, 2003). Methane and CO were measured by the NASA Langley Differential Absorption CO Measurement (DACOM) instrument, which uses two tunable diode lasers in the infrared spectral region to simultaneously measure the absorption of light by CH_4 (3.3 µm) and CO (4.7 µm) (Fried et al., 2008). Nitric oxide, NO_2, NO_y and O_3 were measured using the 4-channel NCAR $NO_{xy}O_3$ chemiluminescence instrument (Weinheimer et al., 1994).

RESULTS AND DISCUSSION

Mixing ratio time series for Flight 23 are shown for selected species in Figs. 4 and 10, and altitudinal profiles are shown for many measured species in Figs. 5–9. Values that are below their LOD have been given a value of "0" so that they are visually represented on the graphs. Measurement statistics for Flight 23 – including the BL excursions over the oil sands mining operations (Leg 7, n=17) and in background air (Leg 9, n = 6) – are given in Table 1. For comparison, statistics for free troposphere (FT) measurements made earlier in the flight are also shown (60–76_ N; n=66). Results for individual samples collected during Leg 7 (Table 2) show that not all of the Leg 7 samples were influenced by emissions from the oil sands because some samples were collected south of the mining operations as the plane manoeuvred (Fig. 2b). For many compounds their maximum enhancements were measured directly downwind of the Syncrude Mildred Lake Facility (i.e., samples 4, 5 and 6 in Table 2 and Fig. 2b). For other compounds such as isoprene,

Table 1: Statistics of boundary layer measurements for 84 compounds measured near oil sands surface mining north of Fort McMurray, Alberta on 10 July 2008 (n=17). Concurrent local background values in the boundary layer (Bkgd) are also included (n=6), as are free tropospheric measurements (FT) collected between 60–76_ N (n = 66). Max. Enh. = Maximum enhancement of oil sands values over background values (oil sands max/bkgd avg) – when the background mixing ratio was below detection limit (LOD) a mixing ratio of 1.5 pptv was assumed in the enhancement calculations; Min = minimum; Max = maximum; Avg = average; StD = standard deviation; n/a = not applicable. Units: pptv unless otherwise stated

Compound	Formula	Lifetime[a]	LOD (pptv)	Precision[b] (%)	Accuracy (%)	Oil Sands Min (pptv)	Max (pptv)	Avg (pptv)	StD (pptv)	BKGD Avg (pptv)	StD (pptv)	FT Avg (pptv)	StD (pptv)	Max. Enh.
Sulphur dioxide	SO_2	1 day	20	12	10	119	38730	4697	11525	102	27	17	5	382
Nitric oxide	NO	10 s	20	2	10	7	4980	635	1403	16	6	3	10	319
Nitrogen dioxide	NO_2	1 day	30	5	10	19	4995	678	19	24	11	8	15	210
Total reactive nitrogen	NO_y	n/a	20	1	10	211	10555	1620	2888	194	33	424	166	54
Ozone	O_3 (ppbv)	8 day	40	1	5	25	31	28	2	31	0	75	23	0.98
Carbon dioxide	CO_2 (ppmv)	> 100 yr	n/a	0.1 ppmv	0.25 ppmv	378	389	381	4	378	1	382	1	1.03
Methane	CH_4 (ppbv)	9 yr	n/a	0.1	2	1844	1983	1876	35	1843[g]	5	1836	10	1.08[g]
Carbon monoxide	CO (ppbv)	2 mo	n/a	1	1	97	144	103	11	97	1	111	13	1.48
Alkanes														
Ethane	C_2H_6	47 day[c]	3	1	5	754	1492	917	174	781	22	813	73	1.9
Propane	C_3H_8	11 day[c]	3	2	3	214	714	382	135	200	26	127	20	3.6
i-Butane	C_4H_{10}	5.5 day[c]	3	3	3	32	290	89	69	29	9	7	4	10
n-Butane	C_4H_{10}	4.9 day[c]	3	3	3	64	765	202	179	63	19	14	7	12
i-Pentane	C_5H_{12}	3.2 day[c]	3	3	3	31	564	141	132	22	10	3	4	26
n-Pentane	C_5H_{12}	3.0 day[c]	3	3	5	27	510	116	119	21	8	LOD	n/a	24
n-Hexane	C_6H_{14}	2.2 day[c]	3	3	5	6	294	44	74	5	3	LOD	n/a	65
n-Heptane	C_7H_{16}	1.7 day[c]	3	3	5	LOD	596	70	158	LOD	n/a	LOD	n/a	397
n-Octane	C_8H_{18}	1.4 day[c]	3	3	5	LOD	363	45	101	LOD	n/a	LOD	n/a	242
n-Nonane	C_9H_{20}	1.2 day[c]	3	3	5	LOD	91	13	25	LOD	n/a	3	3	61
2+3-Methylpentane	C_6H_{14}	2.2 day[c]	3	3	5	11	288	57	71	9	4	LOD	n/a	34
2,3-Dimethylbutane	C_6H_{14}	2.0 day[c]	3	3	5	LOD	28	5	8	LOD	n/a	LOD	n/a	19
Ethene	C_2H_4	1.4 day[c]	3	3	5	23	270	69	69	20	2	4	11	14
Propene	C_3H_6	11 h[c]	3	3	5	7	128	28	30	LOD	n/a	LOD	n/a	85
1-Butene	C_4H_8	8.8 h[c]	3	3	5	LOD	72	13	25	LOD	n/a	LOD	n/a	48
i-Butene	C_4H_8	5.4 h[c]	3	3	5	4	37	9	9	LOD	n/a	LOD	n/a	25
cis-2-Butene	C_4H_8	4.9 h[c]	3	3	5	LOD	9	LOD	n/a	LOD	n/a	LOD	n/a	6
trans-2-Butene	C_4H_8	4.3 h[c]	3	3	5	LOD	18	3	6	LOD	n/a	LOD	n/a	12
1,3-Butadiene	C_4H_6	4.2 h[c]	3	3	5	LOD	8	LOD	n/a	LOD	n/a	LOD	n/a	5
Isoprene	C_5H_8	2.8 h[c]	3	3	5	243	780	468	147	311	95	LOD	n/a	2.5
Alkynes														
Ethyne	C_2H_2	12–17 day	3	3	5	56	138	74	22	59	5	106	34	2.4
Propyne	C_3H_4	2 day	5	30	20	LOD	n/a	LOD	n/a	LOD	n/a	LOD	n/a	n/a
Cycloalkanes														
Cyclopentane	C_5H_{10}	2.3 day[c]	3	3	5	LOD	49	9	10	LOD	n/a	LOD	n/a	27
Methylcyclopentane	C_6H_{12}	2.0 day[c]	3	3	5	4	185	31	47	6	2	3	1	62
Cyclohexane	C_6H_{12}	1.7 day[c]	3	3	5	5	133	23	33	4	1	LOD	n/a	87
Methylcyclohexane	C_7H_{14}	1.3 day[c]	3	3	5	4	339	52	100	3	2	3	2	113
Aromatics														
Benzene	C_6H_6	9.5 day[c]	3	3	5	13	82	24	18	11	1	21	10	7
Toluene	C_7H_8	2.1 day[c]	3	3	5	6	401	50	102	6	1	LOD	n/a	73
Ethylbenzene	C_8H_{10}	1.7 day[c]	3	3	5	LOD	84	8	21	LOD	n/a	LOD	n/a	56
m + p-Xylene	C_8H_{10}	12–19 h[c] [d]	3	3	5	LOD	272	29	74	LOD	n/a	LOD	n/a	181
o-Xylene	C_8H_{10}	20 h[c]	3	3	5	LOD	127	14	37	LOD	n/a	LOD	n/a	85
n-Propylbenzene	C_9H_{12}	2.0 day[c]	3	3	5	LOD	13	2	4	LOD	n/a	LOD	n/a	9
m-Ethyltoluene	C_9H_{12}	15 h[c]	3	3	5	LOD	27	4	9	LOD	n/a	LOD	n/a	18
o-Ethyltoluene	C_9H_{12}	23 h[c]	3	3	5	LOD	17	2	5	LOD	n/a	LOD	n/a	11
p-Ethyltoluene	C_9H_{12}	24 h[c]	3	3	5	LOD	20	2	6	LOD	n/a	LOD	n/a	13
1,2,3-Trimethylbenzene	C_9H_{12}	8.5 h[c]	3	3	5	LOD	40	5	11	LOD	n/a	LOD	n/a	27
1,2,4-Trimethylbenzene	C_9H_{12}	8.5 h[c]	3	3	5	LOD	51	6	15	LOD	n/a	LOD	n/a	34
1,3,5-Trimethylbenzene	C_9H_{12}	4.9 h[c]	3	3	5	LOD	10	1	3	LOD	n/a	LOD	n/a	7
Monoterpenes														
α-Pinene	$C_{10}H_{16}$	5.3 h[c]	3	3	5	15	217	67	52	20	7	LOD	n/a	11
β-Pinene	$C_{10}H_{16}$	3.7 h[c]	3	3	5	38	610	226	149	84	24	LOD	n/a	7

Oxygenated hydrocarbons

Methanol	CH_3OH	12 day[c]	50	30	20	1848	3570	2515	513	1967	354	2276	816	1.8
Ethanol	C_2H_5OH	3.6 day[c]	20	30	20	76[b]	141[b]	106[b]	23[b]	75	12	81	20	1.9[b]
Acetone	C_3H_6O	15 day[e]	100	30	20	393	941	644	154	519	71	947	229	1.8
MEK	C_4H_8O	9.5 day[c]	5	30	20	20	214	65	49	20	16	57	27	11
MAC	C_4H_6O	9.6 h[c]	5	30	20	26	266	92	63	35	10	3	4	8
MVK	C_4H_6O	14 h[c]	5	30	20	42	379	141	109	64	25	16	10	6
MTBE	$C_5H_{12}O$	3.9 day[c]	1	30	20	LOD	n/a	LOD	n/a	LOD	n/a	LOD	n/a	n/a
Furan	C_4H_4O	3.4 h[c]	10	30	20	LOD	n/a	LOD	n/a	LOD	n/a	LOD	n/a	n/a

Halocarbons

CFC-11	CCl_3F	45 yr[f]	10	1	3	251	254	253	1	252.7	0.8	248.4	2.4	1.01
CFC-12	CCl_2F_2	100 yr[f]	10	1	3	526	538	534	3.2	532.3	1	529.2	3.2	1.01
CFC-113	CCl_2FCClF_2	85 yr[f]	5	1	3	77.5	79.5	78.6	0.6	78.1	0.5	78.2	0.7	1.02
CFC-114	$CClF_2CClF_2$	300 yr[f]	1	1	10	16.2	16.8	16.5	0.1	16.4	0.2	16	0.2	1.02
Methyl chloroform	CH_3CCl_3	5.0 yr[f]	0.1	1	5	12.1	12.6	12.3	0.2	12.2	0.1	12.2	0.1	1.04
Carbon tetrachloride	CCl_4	26 yr[f]	1	1	5	91.8	93.4	92.7	0.4	92.3	0.3	91.7	0.9	1.01
Halon-1211	$CBrClF_2$	16 yr[f]	0.1	1	5	4.09	4.37	4.22	0.1	4.16	0.05	4.23	0.07	1.05
Halon-1301	$CBrF_3$	65 yr[f]	0.1	10	10	3.0	3.5	3.3	0.15	3.18	0.12	3.12	0.13	1.10
Halon-2402	$CBrF_2CBrF_2$	20 yr[f]	0.01	1	5	0.51	0.54	0.52	0.01	0.51	0.01	0.51	0.01	1.06
HFC-134a	CH_2FCF_3	14 yr[f]	1	3	10	44	48.9	46.7	1.3	45.3	0.6	46.6	1.6	1.08
HCFC-22	CHF_2Cl	12 yr[f]	2	5	5	189.2	212.1	200	6.7	188.7	1.1	193.4	5.8	1.12
HCFC-141b	CH_3CCl_2F	9.3 yr[f]	0.5	3	10	16.9	21.8	21.2	0.4	20.4	0.7	20.8	0.6	1.07
HCFC-142b	CH_3CClF_2	18 yr[f]	0.5	3	10	18.5	22.1	20.1	1	18.8	0.2	19.5	0.6	1.18
Methyl bromide	CH_3Br	0.7 yr[f]	0.5	5	10	7.6	8.3	8	0.2	7.7	0.1	8.3	0.7	1.08
Methyl chloride	CH_3Cl	1.0 yr[f]	50	5	10	508	545	522	12	503	6	530	13	1.08
Methyl iodide	CH_3I	4 d	0.005	5	20	0.37	0.45	0.41	0.02	0.36	0.01	0.06	0.04	1.26
Dibromomethane	CH_2Br_2	3–4 mo	0.01	5	20	0.76	0.94	0.87	0.04	0.91	0.03	0.75	0.05	1.03
Dichloromethane	CH_2Cl_2	3–5 mo	1	5	10	28.4	35.1	31.3	1.8	28.9	0.6	30.9	1.7	1.22
Chloroform	$CHCl_3$	3–5 mo	0.1	5	10	10.3	15.6	11.7	1.3	10.8	0.3	10.4	0.6	1.45
Trichloroethene	C_2HCl_3	5 d	0.01	5	10	0.1	4.8	0.6	1.2	0.1	0	0.1	0	35.8
Tetrachloroethene	C_2Cl_4	2–3 mo	0.01	5	10	2.7	5.9	3.2	0.8	2.8	0.1	2.9	0.3	2.12
1.2-Dichloroethane	$C_2H_4Cl_2$	1–2 mo	0.1	5	10	7.4	9.4	8.5	0.6	8.1	0.5	8.9	0.8	1.16
Bromodichloromethane	$CHBrCl_2$	2–3 mo	0.01	10	50	0.14	0.17	0.15	0.01	0.15	0.01	0.15	0.01	1.12
Dibromochloromethane	$CHBr_2Cl$	2–3 mo	0.01	5	50	0.09	0.13	0.12	0.01	0.13	0.01	0.11	0.02	1.03
Bromoform	$CHBr_3$	11 mo	0.01	10	20	0.7	0.91	0.83	0.05	0.9	0.04	0.47	0.31	1.01
Ethyl chloride	C_2H_5Cl	1 mo	0.1	5	30	0.75	2.62	1.42	0.47	1.39	0.68	1.67	0.62	1.89

Sulphur Compounds

Carbonyl sulphide	OCS	2.5 yr	10	2	10	392	484	437	26	413	13	445	19	1.17
Dimethyl sulphide	CH_3SCH_3	1–2 d	1	10	20	6	18	10.7	3.9	4.7	0.8	LOD	n/a	3.9

[a]Lifetimes of short-lived OH-controlled compounds are shorter (longer) during the summer (winter), when there are more (fewer) hydroxyl radicals (OH) available for oxidative reactions.

[b]The VOC precision deteriorates as we approach our detection limit; at low values the precision is either the stated precision (%) or 3 pptv, whichever is larger. The NO, NO_2, NO_y and O_3 precision values are for high mixing ratios as were encountered over the oil sands; at low mixing ratios their precision is 20 pptv for the nitrogen species and 0.1 ppbv for O_3.

[c]Based on OH rate constants from Atkinson and Arey (2003) and assuming a 12-h daytime average OH radical concentration of 2.0×10 molec cm^{-3}. The lifetime estimates for furan and methylcyclopentane are from Atkinson et al. (2005) and also use a 12^{-h} daytime OH value of 2.0×10 molec cm^{-3}.

[d]The OH-lifetimes of m-xylene and p-xylene are 12.0 and 19.4 hours, respectively.

[e]Total tropospheric lifetime based on Jacob et al. (2002).

[f]Total lifetimes based on Clerbaux et al. (2007).

[g]CH_4 mixing ratios were not measured during Leg 9, and the background value was determined from the Cold Lake landing (see text).

[h]Ethanol data were not available for the first seven samples of Leg 7, and the statistics presented here are based on samples 8–17.

Which has a dominant biogenic source, the maximum occurred further south over vegetation (i.e., sample 14). Because the Leg 7 samples ranged from near-background air to strongly polluted industrial plumes, the discussion below includes a "maximum enhancement" over the Leg 9 background average, based on each compound's maximum mixing ratio during Leg 7. For correlation purposes, the SO_2, CO_2, CH_4, CO, NO, NO_2, NO_y and O_3 data are based on the average only of those 1 s measurements (30 s for SO_2) that overlapped the VOC sampling times (i.e., the so-called hydrocarbon data merge). The complete 1 s and 30 s data sets for these eight compounds, together with all the VOC data from the summer phase of ARCTAS, are available at ftp://ftp air.larc.nasa.gov/pub/ARCTAS/DC8 AIRCRAFT.

General Features

Atmospheric trace gas mixing ratios are generally greater in the Earth's BL than above in the FT. Here we seek to quantify mixing ratio increases over the oil sands that exceed the increases that may occur as the aircraft descends into the boundary layer. The 84 trace gases presented here can be classified into three groups: (1) compounds that were strongly enhanced (>10%) over the oil sands relative to levels in the local background BL air; (2) compounds that showed minimal increases over the oil sands (<10% greater than the local background average); and (3) compounds that showed no statistical enhancements over the oil sands compared to the local background. The first group, easily the largest, includes SO_2, NO, NO_2, NO_y, CO_2, CH_4, CO and 70% of the measured VOCs, namely alkanes, aromatics, cycloalkanes, alkenes, oxygenated hydrocarbons, ethyne, short-lived solvents (C_2Cl_4, C_2HCl_3, $CHCl_3$, CH_2Cl_2) and some HCFCs (HCFC-22 and HCFC-142b) (Sects. 3.2, 3.3.3 and 3.4–3.6). The maximum enhancements were 1.1–397× the background BL values, most notably nheptane (397×), SO_2 (382×), NO (319×), n-octane (242×), NO_2 (210×), m+p-xylene (181×) and methylcyclohexane (113×). Interestingly this group comprises _-pinene and _- pinene, which are usually associated with biogenic emissions (Sect. 3.2.3). In the second group, halocarbons including HFC-134a, HCFC-141b, the halons and the methyl halides were minimally enhanced over the oil sands (Sect. 3.3.2). In the third group, long-lived industrial halocarbons (e.g., CFCs, CCl_4, CH_3CCl_3), several brominated species (e.g., $CHBr_3$, CH_2Br_2) and ozone (O_3) were not enhanced over the oil sands (Sects. 3.3.1 and 3.5). In fact O_3 was relatively depleted because of titration by high levels of NO.

Table 2: Mixing ratios of selected compounds in 17 whole air samples measured near the Alberta oil sands on 10 July 2008 during Flight Leg 7. Samples 4, 5 and 6 were collected directly downwind of the Syncrude Mildred Lake Facility and showed the strongest enhancements for most compounds. An exception is the biogenic tracer isoprene, which was most enhanced in sample 14. Maximum mixing ratios are shown in bold for each compound. To help distinguish which of the 17 samples collected during Leg 7 resembled background air as the aircraft manoeuvred south of the mining operations, the bottom two rows show the average (standard deviation) values for the background samples collected during Flight Leg 9 (n = 6). Lat = latitude; Long = longitude; Alt = altitude; WD = wind direction; Bkgd = background; StD = standard deviation; n/a = not applicable; LOD denotes values below the detection limit

No.	Time	Lat	Long	Alt	WD	NO_y	SO_2	C_2Cl_4	Ethane	Ethene	α-Pinene	n-Hexane	n-Heptane	Cyclo-Hexane	Toluene	o-Xylene	Isoprene
1	11:05	56.887	248.822	854	207	517	190	3.39	844	54	72	16	7	10	16	LOD	457
2	11:06	56.956	248.773	838	201	889	264	2.98	833	42	79	15	5	11	13	LOD	391
3	11:07	57.027	248.714	832	207	1461	6082	3.71	884	54	130	17	6	12	20	4	378
4	11:08	57.079	248.608	761	190	10554	38730	5.93	1492	270	217	141	268	47	135	94	507
5	11:09	57.118	248.487	747	193	7343	30976	3.77	1086	214	138	294	596	133	401	127	427
6	11:10	57.148	248.356	721	184	2795	1766	3.17	1106	89	31	90	236	70	164	18	399
7	11:11	57.123	248.236	774	225	699	299	3.22	827	95	43	13	5	12	8	LOD	344
8	11:13	56.979	248.162	794	243	259	208	2.81	794	30	17	11	4	10	9	LOD	344
9	11:15	56.905	248.174	815	241	219	190	2.74	784	23	39	6	3	6	8	LOD	363
10	11:17	56.845	248.257	846	228	231	140	2.82	809	28	15	7	5	6	8	LOD	466
11	11:20	56.785	248.348	842	225	236	119	2.79	881	42	63	10	5	8	8	LOD	768
12	11:22	56.725	248.444	843	230	211	130	2.76	797	27	42	6	3	6	6	LOD	466
13	11:24	56.665	248.532	802	234	240	130	2.71	850	31	46	10	LOD	5	8	LOD	677
14	11:26	56.607	248.620	787	224	296	154	2.77	902	38	55	16	7	9	9	LOD	780
15	11:29	56.574	248.741	785	202	645	157	2.90	940	47	48	41	14	19	12	LOD	455
16	11:32	56.564	248.876	804	194	579	174	2.79	944	50	68	37	13	16	12	LOD	496
17	11:34	56.601	248.989	817	194	373	135	2.82	874	37	28	17	7	10	8	LOD	243
Bkgd	11:57	55.9	246.6	1135	270	194	102	2.8	781	20	20	5	LOD	4	6	LOD	311
(StD)	(0.06)	(0.3)	(0.5)	(180)	(26)	(33)	(27)	(0.1)	(22)	(2)	(7)	(3)	(n/a)	(1)	(1)	(n/a)	(95)

Before discussing the results below, it is important to first recognize the limitations of this data set. Because of its short time-frame (17 min) and sample size (n=17), the representativeness of the measured VOC enhancements is unclear. On the other hand, because emissions from oil sands mining are so poorly characterized in the peer-reviewed literature due to the inaccessibility of mining sites to independent observers, this study provides important close-range observations of the VOCs that are being released, i.e., which individual species are emitted and with what other species do they correlate. The large number of compounds and relatively small number of samples in this data set make it unsuitable for emission estimates or factor analysis such as Principal Component Analysis or Positive Matrix Factorization (e.g., Thurston and Spengler, 1985; Paatero, 1997; Choi et al., 2003), and we have instead performed linear correlations among the measured compounds using least squares fits to better understand their source influences.

Non-Methane Volatile Organic Compounds (NMVOCs)

Non-methane volatile organic compounds (NMVOCs) are mainly emitted from three major sources: vegetation, biomass burning, and anthropogenic activities such as industry and fossil fuel production, distribution and combustion (e.g., Guenther et al., 2000; Ehhalt and Prather, 2001; Folberth et al., 2006). Remarkably, of the 48 C_2–C_{10} hydrocarbons that we measured, all but three (propyne, furan and methyl tert-butyl ether or MTBE) showed very strong enhancements over the oil sands (maximum values at least 80% more than the local background average). Based on their correlations with one another (Table 3), the NMVOCs appear to have two distinct sources: (1) the oil sands and its products and/or the diluent used to lower the viscosity of the extracted bitumen (enhanced levels of C_4–C_9 alkanes, C_5–C_6 cycloalkanes, C_6–C_8 aromatics), and (2) industrial sources that support the mining effort, such as upgrading (elevated levels of C_2–C_4 alkanes, C_2–C_4 alkenes, C_9 aromatics). The major upgrading processes at this location are distillation (to separate different hydrocarbons), thermal conversion/coking (to convert the bitumen into lighter, refinable hydrocarbons), catalytic conversion (an enhanced form of thermal conversion), and hydrotreating (the addition of hydrogen to unsaturated molecules to stabilize them). For example the catalytic hydrocracking of Athabasca bitumen vacuum bottoms (ABVBs) produces gaseous by-products including hydrogen sulphide (H_2S), C_1–C_7 alkanes and, to a lesser extent, C_2–C_4 alkenes (Dehkissia et al., 2004)

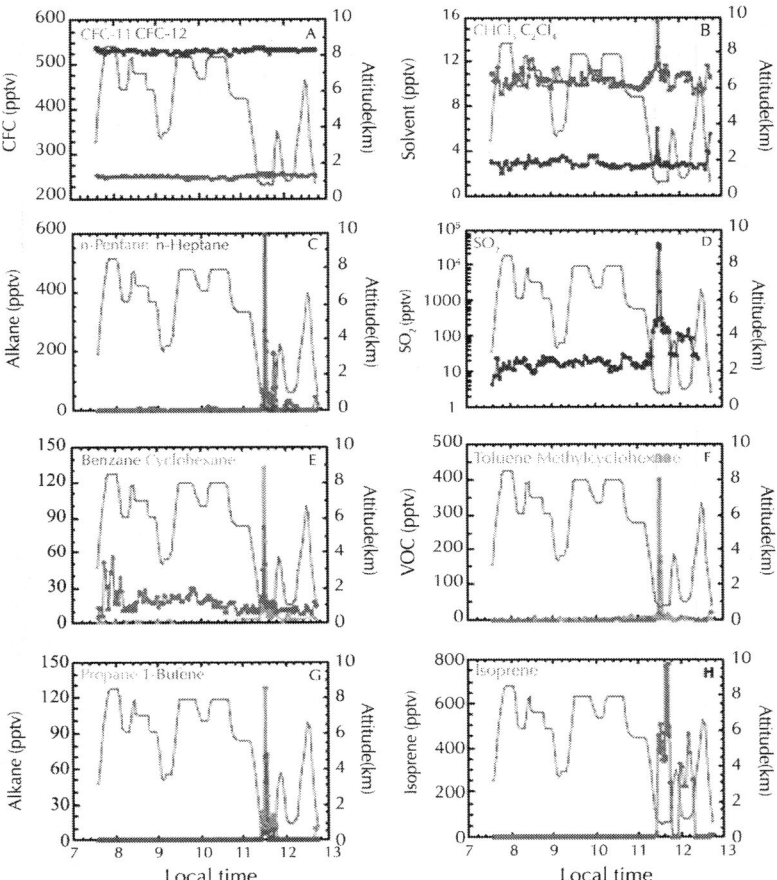

Figure 4: Time series of selected trace gases measured during ARCTAS Flight 23 on 10 July 2008. The solid blue line traces altitude. The boundary layer run over oil sands mining operations (Leg 7, shaded) is followed by a second boundary run (Leg 9) in cleaner air 1_ further south. Values below detection limit have been given a value of "0" for plotting purposes. Note that the mixing ratios of n-heptane, the cycloalkanes, the butenes and isoprene were below their detection limits for most of the flight.

Alkanes, Cycloalkanes and Aromatics

Because of increasing reactivity with the hydroxyl radical (OH) with increasing chain length, the average (±1σ) nalkane mixing ratios decreased with increasing chain length in the background BL (Leg 9), from ethane (781±22 pptv) to propane (200±26 pptv) to n-butane (63±19 pptv), etc. (Table 1). The $_-C_7$ n-alkanes are so short-lived (<2 days) that they were undetectable in the background BL (<3 pptv). By contrast, all 12 C_2–C_9 alkanes were strongly enhanced in the oil sands plumes. Maximum ethane (1490 pptv) and propane (714 pptv) levels occurred in sample 4 downwind of the Syncrude Mildred Lake Facility (Fig. 2b) and were almost double and quadruple the background BL values, respectively (Table 1; Fig. 5a–b). Despite minimal urbanization in northern Alberta, the oil sand plume values were greater than average summertime values measured in rural New England from 2004–2007 (Russo et al., 2010). Instead they were comparable to average ethane and propane levels measured in urban areas of Baltimore and New York City during a summer study of 28 US cities from 1999– 2005 (Baker et al., 2008). By comparison, the ethane and propane values in the oil sands plumes were smaller than median values measured during summer 2006 in the Houston and Galveston Bay area, which is both a large urban area and a major petrochemical manufacturing center (4407 pptv ethane; 2713 pptv propane) (Gilman et al., 2009). The primary ethane and propane sources are fossil fuel production (mainly unburned gas in the case of ethane) and biofuel and biomass burning, plus propane is also emitted in gasoline exhaust (e.g., Watson et al., 2001; Buzcu and Fraser, 2006; Xiao et al., 2008). Ethane and propane correlated well with a range of industrial and combustion species such as other light alkanes, alkenes, NO, NO_2, NO_y and ethyne (Table 3). By contrast, even though ethane is the second most abundant component of natural gas after CH_4 (Xiao et al., 2008; McTaggart-Cowan et al., 2010), it did not correlate well with CH_4 (r^2 =0.52). This relatively poor correlation is surprising given the heavy use of natural gas during bitumen extraction (Sect. 1) and appears to indicate low natural gas leakage levels (see Sect. 3.6). Because ethane and propane are primarily associated with evaporative rather than combustive fossil fuel emissions, we expect a co-located evaporative source from the industries that process the oil sands, possibly from fuel gas and/or hydrocracking.

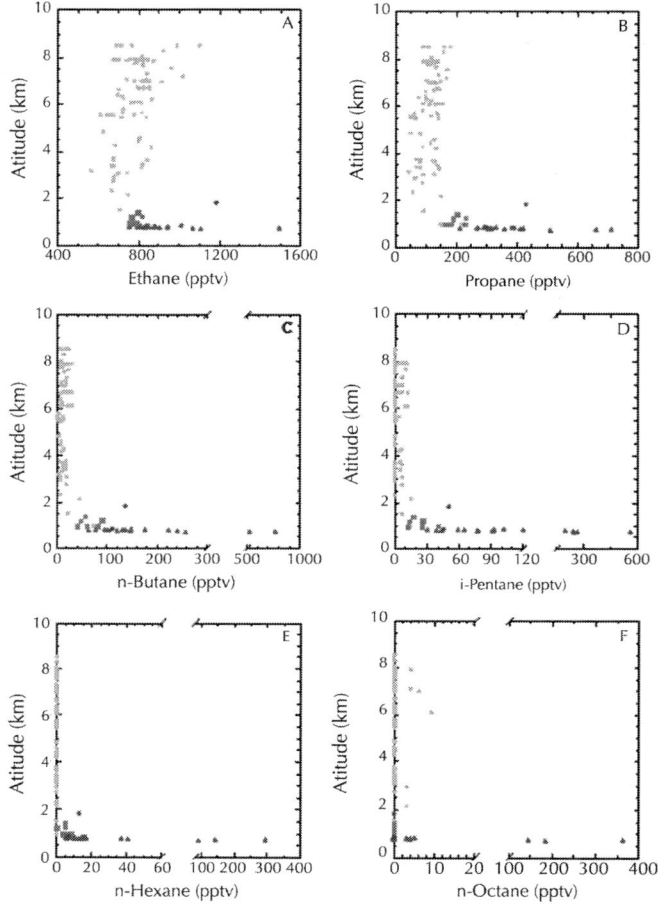

Figure 5: Vertical profiles of selected alkanes measured during Flight 23 on 10 July 2008. Red triangles: oil sands boundary layer run (Leg 7); green squares: background boundary layer run (Leg 9); purple diamonds: Cold Lake landing; blue circles: remainder of Flight 23. Note that the mixing ratios of n-hexane and n-octane were below their detection limits for most of the flight.

The butanes are associated with fossil fuel evaporation (e.g., liquefied petroleum gas or LPG) and biomass burning (Blake et al., 1996; Watson et al., 2001; Buzcu and Fraser, 2006). The i-butane (290 pptv) and n-butane (765 pptv) maxima in the oil sands plumes were 10–12× greater than in the background BL (Table 1; Fig. 5c) and were comparable to maximum levels measured in fresh biomass burning plumes during ARCTAS (not shown). They were also similar to the lower range of values measured in the 28 US city study (Baker et al., 2008), and they were much greater than average July values measured in rural New England (70±39 pptv and 105±59 pptv, respectively; Russo et al., 2010), even though there is widespread LPG leakage in the northeastern United States (White et al., 2008; Russo et al., 2010). The ratio of i-butane/n-butane varies according to source, for example 0.2–0.3 for vehicular exhaust, 0.46 for LPG, and 0.6–1.0 for natural gas (Russo et al., 2010 and references therein). Interestingly, the i-butane/n-butane ratio for the dominant oil sands plumes, i.e., samples 4, 5 and 6 (0.42±0.03 pptv pptv^{-1}) was very close to that expected for LPG. We have not been able to determine that LPG is used in the mining operations, and this signal could potentially represent emissions from hydrocrackers and/or the fuel gas that is produced and used at the major oil sands operations. Even though they only contain 4 carbon atoms, the butanes correlated better with the suite of heavier compounds that is associated with direct emissions from the oil sands and/or diluent (e.g., n-hexane, xylenes, cyclopentane; 0.93_ r^2 _0.96) than with species linked to industrial activities associated with the mining operations (e.g., propane, $CHCl_3$, ethene, ethyne, NO, NO_y; 0.77_r_2 _0.84) (Table 3). Because the butanes are not generally associated with combustive fossil fuel emissions, these results indicate multiple evaporative butane sources at the mining sites.

Crude oil is composed of $\geq C_5$ alkanes, aromatics, cycloalkanes and asphaltics (Sect. 1), and even though the oil sands are extra-heavy and include molecules with 50 or more carbon atoms, they are still expected to contain a certain fraction of volatiles that would be captured by our measurements (D. Spink, personal communication, 2010). The C_5– C_9 n-alkanes showed very strong enhancements, with higher values over the oil sands than at any other time during the summer phase of ARCTAS. It is noteworthy that the maximum abundances were not ordered according to chain length. n-Heptane (596 pptv) showed the highest maximum mixing ratio, followed by the pentanes (510–560 pptv), n-octane (363 pptv) and n-hexane (294 pptv) (Table 1; Figs. 4c, 5d– f). These values represent enhancements of 24–397× the background BL values (in the enhancement calculations a background value of 1.5 pptv was assumed for BL measurements that were below our detection limit of 3 pptv). Oil having a plant origin is known to show a predominance for n-alkanes with an odd chain number, in contrast to n-alkanes that are produced from marine organic matter (Rogers and Koons, 1971). Because oil sands are derived from marine phytoplankton, the relative abundance of pentanes and nheptane may suggest a stronger signal from the diluent than from the oil sands, assuming the diluent is made from conventional plant-based oil. Branched alkanes such as 2,3- dimethylbutane and the methylpentanes (which we present as a single measurement, i.e., 2+3-methylpentane) were also enhanced by factors of 19–34 over the local background (Table 1).

The cycloalkanes were close to or below our 3 pptv detection limit during Leg 9, as they had been in the FT during the transit from Thule. By contrast, all the cycloalkanes were strongly enhanced over the oil sands, with maximum mixing ratios ranging from 41 pptv (cyclopentane) to 339 pptv (methylcyclohexane) (Table 1; Figs. 4e, 6a). Cycloalkanes are a major component of crude oil (Sect. 1), and these results are consistent with their evaporative release from oil sands surface mining operations.

Aromatics are associated with combustion (fossil fuel and biomass burning), fuel evaporation, industry, and biogenic emissions (e.g., Karl et al., 2009a; White et al., 2009). With the exception of the longest-lived aromatics that we measured – i.e., benzene (1–2 weeks) and toluene (2–3 days) – none of the aromatics were detectable in either the FT or in background BL air, whereas they were all strongly elevated over the oil sands with maximum enhancements of 7–181× over the background BL (Table 1; Figs. 4e–f, 6b–d). Toluene was the most abundant aromatic in the oil sands plumes (401 pptv maximum) followed by the xylenes (127–272 pptv maxima) and benzene (82 pptv maximum). For comparison these aromatic enhancements are much smaller than the maximum values measured near major petrochemical complexes in urban/industrial areas of Houston, Texas and southern Catalonia, Spain (e.g., toluene maxima of 16 000–77 000 pptv; Gilman et al., 2009; Ras et al., 2009). That is, the release of aromatics from the oil sands mining sites appears to be much smaller than releases further downstream during petrochemical refining.

Table 3: Correlation matrix for selected VOC species measured over oil sands mining operations in Alberta on July 10, 2008 (n = 17). Correlations ≥ 0.7 are highlighted in bold. With the exception of $CHCl_3$ and C_2Cl_4, all the halocarbons showed $0.00 \leq r^2 \leq 0.43$ with the selected species and are not included

	SO_2	NO	Ethene	Propane	n-Butane	n-Heptane	Toluene	Methyl cyclohexane
SO_2	1.00	0.97	0.92	0.76	0.79	0.66	0.57	0.43
NO	0.97	1.00	0.95	0.82	0.77	0.68	0.57	0.46
NO_2	0.96	1.00	0.95	0.83	0.77	0.67	0.57	0.43
NO_y	0.96	1.00	0.96	0.83	0.77	0.67	0.57	0.46
O_3	0.39	0.44	0.51	0.68	0.54	0.38	0.35	0.32
CO_2	0.64	0.70	0.74	0.62	0.66	0.70	0.66	0.60
CH_4	0.24	0.37	0.39	0.57	0.39	0.48	0.46	0.53
CO	0.63	0.57	0.60	0.59	0.90	0.92	0.93	0.85
Ethane	0.69	0.81	0.77	0.84	0.54	0.41	0.32	0.26
Propane	0.76	0.82	0.82	1.00	0.84	0.69	0.61	0.55
i-Butane	0.81	0.81	0.82	0.88	0.99	0.86	0.80	0.71
n-Butane	0.79	0.77	0.79	0.84	1.00	0.89	0.85	0.76
i-Pentane	0.54	0.54	0.57	0.72	0.92	0.85	0.85	0.82
n-Pentane	0.55	0.55	0.57	0.70	0.93	0.88	0.88	0.85
n-Hexane	0.70	0.69	0.70	0.73	0.96	0.98	0.96	0.89
n-Heptane	0.66	0.68	0.67	0.69	0.89	1.00	0.99	0.94
n-Octane	0.57	0.61	0.61	0.65	0.83	0.99	0.99	0.97
n-Nonane	0.72	0.74	0.72	0.72	0.89	0.99	0.96	0.90
2+3-Methylpentane	0.65	0.66	0.68	0.77	0.95	0.96	0.94	0.90
2,3-Dimethylbutane	0.74	0.78	0.77	0.84	0.93	0.90	0.86	0.80
Ethene	0.92	0.95	1.00	0.82	0.79	0.67	0.58	0.46
Propene	0.87	0.93	0.95	0.75	0.61	0.48	0.38	0.28
1-Butene	0.54	0.70	0.67	0.73	0.55	0.63	0.56	0.56
i-Butene	0.85	0.90	0.94	0.71	0.62	0.52	0.43	0.32
c-2-Butene	0.91	0.93	0.87	0.67	0.58	0.43	0.33	0.22
t-2-Butene	0.69	0.83	0.78	0.79	0.65	0.71	0.63	0.60
1,3-Butadiene	0.58	0.64	0.57	0.40	0.21	0.10	0.05	0.01
Isoprene	0.00	0.00	0.00	0.01	0.00	0.01	0.01	0.01
Ethyne	0.90	0.92	0.96	0.83	0.79	0.63	0.54	0.43
Cyclopentane	0.61	0.62	0.61	0.73	0.93	0.94	0.93	0.90
Methylcyclopentane	0.57	0.58	0.60	0.67	0.89	0.98	0.99	0.97
Cyclohexane	0.51	0.53	0.55	0.63	0.83	0.97	0.98	0.99
Methylcyclohexane	0.43	0.46	0.46	0.55	0.76	0.94	0.98	1.00
Benzene	0.60	0.64	0.64	0.70	0.86	0.97	0.97	0.95
Toluene	0.57	0.57	0.58	0.61	0.85	0.99	1.00	0.98
Ethylbenzene	0.71	0.66	0.66	0.61	0.91	0.95	0.93	0.84
m+p-Xylene	0.82	0.79	0.79	0.71	0.94	0.95	0.91	0.80
o-Xylene	0.91	0.88	0.86	0.75	0.93	0.89	0.82	0.70
n-Propylbenzene	0.92	0.92	0.89	0.79	0.91	0.88	0.81	0.69
m-Ethyltoluene	0.97	0.96	0.95	0.79	0.86	0.78	0.69	0.56
o-Ethyltoluene	0.98	0.95	0.91	0.73	0.82	0.70	0.60	0.46
p-Ethyltoluene	0.90	0.90	0.88	0.79	0.92	0.90	0.83	0.72
1,2,3-Trimethylbenzene	0.99	0.99	0.94	0.79	0.76	0.65	0.55	0.43
1,2,4-Trimethylbenzene	0.99	0.98	0.95	0.79	0.82	0.73	0.63	0.50
1,3,5-Trimethylbenzene	0.98	0.96	0.92	0.73	0.74	0.61	0.50	0.37

α-Pinene		0.76	0.71	0.69	0.61	0.52	0.32	0.26	0.16
β-Pinene		0.64	0.57	0.57	0.56	0.49	0.26	0.21	0.12

Methanol	0.41	0.52	0.50	0.58	0.40	0.46	0.42	0.41
Ethanol	0.21	0.25	0.31	0.18	0.18	0.22	0.22	0.20
Acetone	0.17	0.24	0.38	0.36	0.29	0.33	0.32	0.34
MEK	0.76	0.85	0.85	0.81	0.62	0.52	0.43	0.36
MAC	0.27	0.31	0.50	0.31	0.25	0.22	0.19	0.16
MVK	0.43	0.44	0.60	0.42	0.35	0.29	0.25	0.20
$CHCl_3$	0.52	0.54	0.60	0.66	0.80	0.85	0.86	0.85
C_2Cl_4	0.78	0.81	0.79	0.60	0.43	0.29	0.21	0.13
OCS	0.20	0.19	0.21	0.06	0.03	0.05	0.04	0.02
DMS	0.50	0.56	0.63	0.54	0.46	0.50	0.47	0.44

With the exception of the C_9 aromatics, the $\geq C_5$ alkanes, cycloalkanes and aromatics showed excellent mutual correlations (Table 3). Interestingly, the best correlations were not limited to compounds within the same class but were mixed among the compound classes. For example, n-heptane, the most strongly enhanced VOC, showed top 10 correlations with n-nonane, n-octane, toluene, methylcyclopentane, nhexane, benzene, cyclohexane, 2,3-methylpentane, m+pxylene, and ethylbenzene ($0.95 \leq r^2 \leq 0.99$). These results show the fugitive co-emissions of a wide range of C_5–C_9 VOCs from oil sands surface mining sites in Alberta, e.g. from oil sands and/or diluent. By contrast, the C_9 aromatics (i.e., ethyltoluenes, trimethylbenzenes) correlated best with each other and with combustion and industrial tracers such as NO, NO_2, NO_y, ethene, propene and cis-2-butene, suggesting their release from industries at the mining sites as opposed to direct evaporative release from the oil sands or diluent.

Alkenes and Alkynes

Because they are primarily biogenic tracers, isoprene and the monoterpenes are considered separately below (Sect. 3.2.3). The 7 C_2–C_4 alkenes considered here are highly reactive, short-lived compounds that are primarily associated with industrial emissions and incomplete combustion (e.g., Sprengnether et al., 2002; Buzcu and Fraser, 2006; de Gouw et al., 2009). Whereas only ethene was detectable in the background BL during Flight 23 (20±2 pptv), all of the alkenes were strongly enhanced over the oil sands by factors of up to 5–85 (Table 1; Figs. 4g, 6e). The maximum mixing ratio decreased with increasing chain length, from ethene (270 pptv) to propene (128 pptv) to i-butene (37 pptv), etc. Although these oil sands values are much greater than the rural background, they are smaller for example than in fresh biomass burning plumes encountered during ARCTAS (e.g., ethene up to 18 690 pptv, propene up to 5465 pptv). They are 260–2430 pptv ethene; 68–500 pptv propene; 33–1550 pptv i-butene; Baker et al., 2008). In the Houston area – where the greatest concentration of petrochemical facilities in the US is located and ethene and propene play a major role in rapid O_3 formation (Ryerson et al., 2003) – average respective ethene and propene mixing ratios of 2690 pptv and 1540 pptv were measured in the summer of 2006 (Gilman et al., 2009), and poor correlation between ethene and CO suggested the dominance of industrial point sources (de Gouw et al., 2009). Over the oil sands, ethene showed excellent correlation with the combustion tracer ethyne ($r^2 = 0.96$), some correlation with CO ($r^2 = 0.60$), and good correlation with the industrial tracer C_2Cl_4 ($r^2 = 0.79$) (Table 3), suggesting the mixed influence of industrial and combustion sources on ethene and other alkenes at the oil sands mining sites.

Propyne was not detectable in any of the samples that we collected during Flight 23. Ethyne – a tracer of incomplete combustion by biomass burning and urban fossil fuel (e.g., Blake et al., 2003; Warneke et al., 2007) – correlated most strongly with ethene, NO_y, NO and the trimethylbenzenes ($0.91 \leq r^2 \leq 0.96$) and showed a maximum mixing ratio of 138 pptv over the oil sands, compared to 59±5 pptv in the local background (Table 1; Fig. 6f). As with the alkenes, this enhancement is small compared to average levels measured in Houston (473 pptv; Gilman et al. 2009) and in the 28 US city study (260–2390 pptv; Baker et al., 2008), indicating the relatively low impact of industrial combustion on the measured ethyne levels.

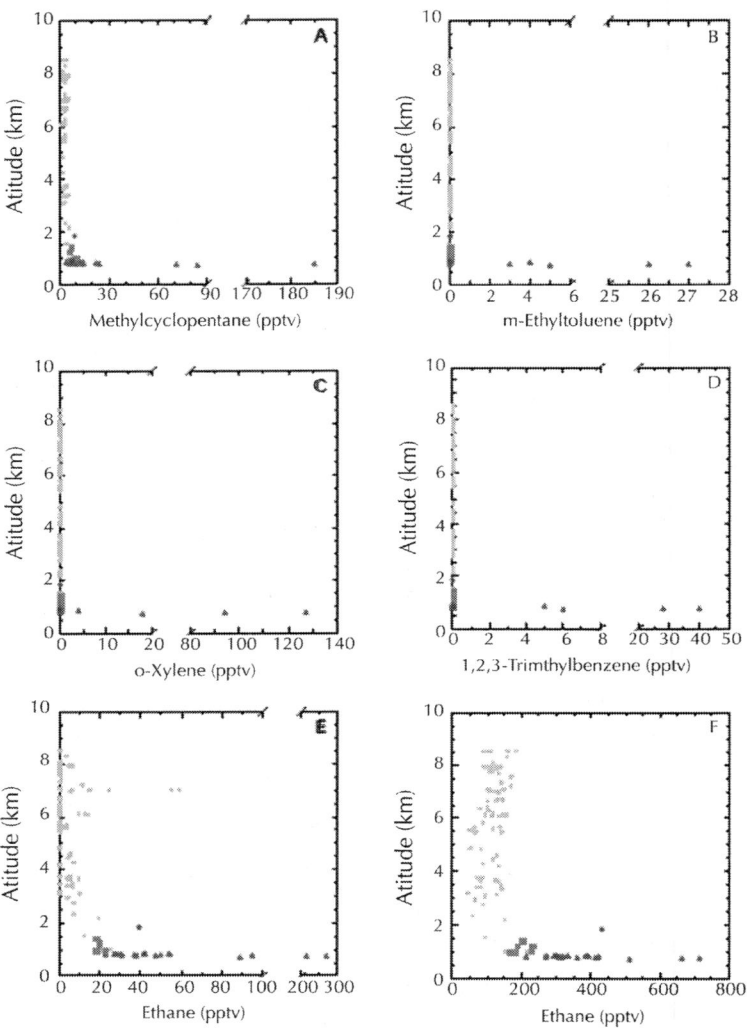

Figure 6: As in Fig. 5 but for methylcyclopentane, selected aromatics, ethene and ethyne. Note that the mixing ratios of m-ethyltoluene, oxylene and 1,2,3-trimethylbenzene were below their detection limits for most of the flight.

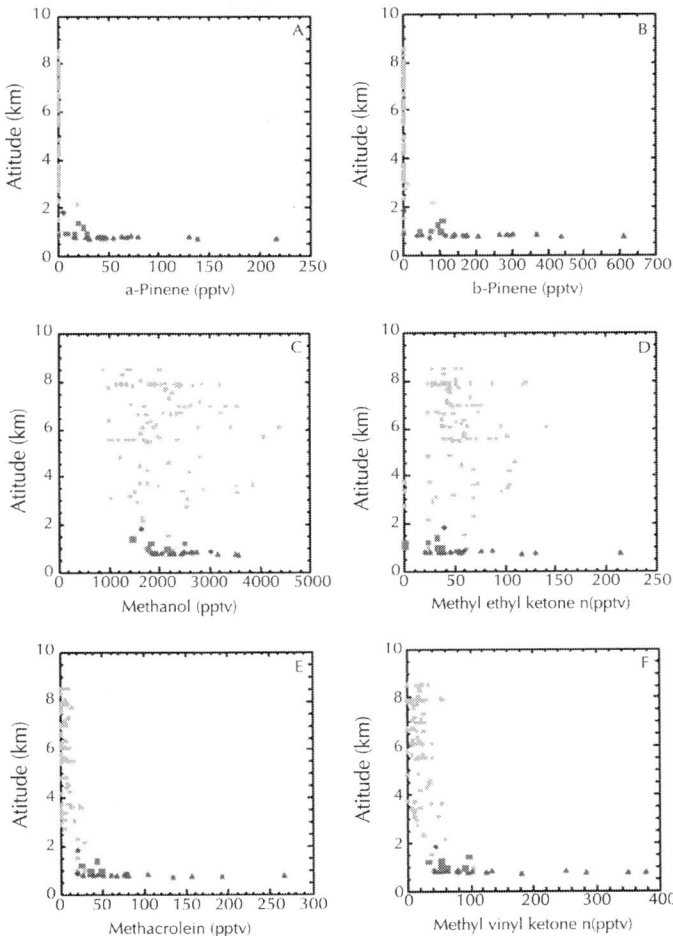

Figure 7: As in Fig. 5 but for the pinenes and selected oxygenated hydrocarbons. Note that the mixing ratios of α-pinene and β-pinene were below their detection limits for most of the flight.

Isoprene and Monoterpenes

Isoprene (C_5H_8) is a short-lived (~ 2.8 h) biogenic volatile organic compound (BVOC) with a major terrestrial plant source (e.g., Sharkey et al., 2008; Warneke et al., 2010). Isoprene levels vary widely in North America, with greatest emission rates in the eastern US during summer

(Guenther et al., 2006). For example in the 28 US city study the lowest average isoprene mixing ratio was measured in El Paso, Texas (47 ± 75 pptv) and the highest in Providence, Rhode Island (2590 ± 1610 pptv) (Baker et al., 2008). In our study, isoprene mixing ratios exceeded 200 pptv in all 17 air samples collected over the oil sands as well as in the background BL air. Interestingly, the average isoprene mixing ratio was 50% greater during Leg 7 (468 ± 147 pptv) than in the local background air (311 ± 95 pptv), with a maximum value of 780 pptv (Tables 1 and 2; Fig. 4h). Anthropogenic sources of isoprene include traffic (e.g., Reimann et al., 2000; Borbon et al., 2001; Barletta et al., 2002) and industry (e.g., Ras et al., 2009). However the isoprene maximum was measured in sample 14, south of the main surface mining operations (Fig. 2b). Isoprene was poorly correlated with all of the trace gases that we measured ($r^2 <0.37$) and its enhancements appear to be biogenic rather than industrial in origin. We believe that the stronger isoprene enhancements during Leg 7 than Leg 9 most likely reflect natural source strength variability rather than an altitude effect. Isoprene is highly reac-tive with OH and it is the shortest-lived species that we measure (Table 1). Although the Leg 7 samples were collected at a lower average altitude (804 ± 38 m) than the background BL measurements (1135 ± 180 m) (Sect. 2) – and in principle more isoprene oxidation by OH could have occurred during transport to higher altitudes during Leg 9 – the individual measurements did not show a clear declining trend with altitude. Instead we expect that samples with higher isoprene levels represent parts of Leg 7 that were downwind of more, or more recent, isoprene emissions. For example aspen are a major isoprene emitter (Sharkey et al., 2008) and the forests in the Fort McMurray area include a mix of trembling aspen and white spruce.

Like isoprene, monoterpenes ($C_{10}H_{16}$) are short-lived (~ 3.7–5.3 h) BVOCs that are emitted by vegetation, though in this case they are more strongly emitted by coniferous ecosystems than by temperate deciduous forests (Fuentes et al., 2000). During this study the pinenes were enhanced during both BL runs compared to measurements in the FT (Fig. 7a–b). Interestingly, however, the maximum α-pinene and β-pinene mixing ratios over the oil sands (217 pptv and 610 pptv, respectively) were much greater (7–11x) than their respective average values in the background BL air (20 ± 7 and 84 ± 24 pptv; Table 1, Fig. 7a–b). At first glance this appears to be consistent with a number of studies that have shown increased emissions of monoterpenes and other VOCs in response to various plant stress factors (Schade and Goldstein, 2003; R̈ais̈anen et al., 2008; Holopainen and Gershenzon, 2010; Niinemets, 2010). For example, monoterpene mixing ratios and emissions from a California ponderosa pine forest showed a 10–30 fold increase during and after major forest thinning, with measured mixing ratios exceeding 3000 pptv (Schade and Goldstein, 2003). However, unlike isoprene, the pinenes were most strongly enhanced in sample 4, downwind of the Syncrude Mildred Lake Facility (Table 2; Fig. 2b), and they showed strong correlations with many species that were elevated in the oil sands plumes. For example α-pinene correlated best with β-pinene (r^2 =0.92) and industrial compounds such as C_2Cl_4 (r^2 = 0.81), SO_2 (r^2 =0.76), 1,2,3-TMB (r^2 =0.74), i-butene (r^2 =0.73) and ethyne (r^2 =0.73). The land surrounding the upgraders is already disturbed and we do not expect the industrial upgrader plumes to have had the opportunity to mix with plumes representing damaged vegetation. Instead these unexpected observations lead us to infer that there could be a source of pinenes associated with the oil sands mining industry, which requires further investigation.

Oxygenated Hydrocarbons

During ARCTAS UC-Irvine measured methanol (CH_3OH), ethanol (C_2H_5OH), acetone (C_3H_6O), methyl ethyl ketone (MEK, C_4H_8O), methacrolein (MAC, C_4H_6O), methyl vinyl ketone (MVK, C_4H_6O), MTBE ($C_5H_{12}O$), and furan (C_4H_4O). The sources of oxygenated hydrocarbons are both natural and anthropogenic and include vegetation, biomass burning, atmospheric production, the oceans and industry (e.g., Horowitz et al., 2003; Jacob et al., 2005; Folberth et al., 2006; Jordan et al., 2009).

Methanol has a major biogenic source and minor sources including biomass burning and anthropogenic emissions (e.g., vehicles, solvent use and manufacturing; Jacob et al., 2005 and references therein). Its maximum mixing ratio (3570 pptv) was measured downwind of the oil sands operations in sample 6, which was enhanced by a factor of 1.8 over the local background average (Table 1; Fig. 7c). Methanol correlated best with MEK and the butenes ($0.71 \leq r^2 \leq 0.75$) suggesting its industrial release from the oil sands facilities. However the oil sands industry appears to be a relatively minor methanol source. For example, the mixing ratios in the oil sands plumes were comparable to those measured at higher altitudes and latitudes during earlier portions of Flight 23, and they were an order of magnitude smaller than maximum methanol values that were measured in fresh biomass burning plumes during ARCTAS (32 740 pptv).

In addition to being produced by plants and used as a solvent and chemical feedstock, ethanol is increasingly being used as a motor fuel and fuel additive (e.g., Kesselmeier and Staudt, 1999; Russo et al., 2010). The maximum ethanol mixing ratio during Leg 7 (141 pptv) was 1.9× the background average, and ethanol correlated best with CH^3CCl^3 and the halons ($0.65 \leq r^2 \leq 0.83$). However, unlike the other gases that we measured, ethanol data were not available for the first seven samples of Leg 7 and therefore we are unable to fully characterize its emissions from the oil sands industry.

Acetone is produced by vegetation, the oceans and the atmospheric oxidation of C_3–C_5 isoalkanes, especially propane (Jacob et al., 2002). It is used industrially as a solvent and polymer precursor and has been observed in abundance downwind of petrochemical complexes (Cetin et al., 2003). The maximum acetone mixing ratio over the oil sands (941 pptv) was 1.8× its local background average. Like methanol, the acetone levels over the oil sands were comparable to other times during the flight and were smaller than maximum values measured in fresh biomass burning plumes during ARCTAS (4552 pptv). Acetone correlated best with MAC, 1,2-dichloroethane and CH_2Cl_2 ($0.64 \leq r^2 \leq 0.69$) suggesting that it is emitted by the oil sands industry but in smaller amounts than other regional sources such as biomass burning.

Methyl ethyl ketone is produced industrially and used as a solvent, and it is known for its characteristic sweet odour (Kabir and Kim, 2010). The maximum MEK mixing ratio over the oil sands (214 pptv) occurred in sample 4 and was 11× greater than the local background average. However most of the MEK enhancements over the oil sands were comparable to those measured at other times during the flight (Fig. 7d). Methyl ethyl ketone showed strong correlations ($r^2 > 0.8$) with a dozen industrial and combustion tracers including C_2–C_4 alkenes, C_2Cl_4, NO and 1,2,3-TMB, showing its release from the oil sands industry in what appear to be relatively minor quantities.

Methacrolein and MVK are major isoprene oxidation products (e.g., Montzka et al., 1993; Stroud et al., 2001; Karl et al., 2009b). The MAC and MVK mixing ratios were greater over the oil sands than at any other time during the flight (Fig. 7e–f), with maximum respective values of 266 and 379 pptv in sample 7, or 8× and 6× the local background average (Table 1). Methacrolein and MVK correlated most strongly with each other (r^2 = 0.87), followed by species such as DMS and 1,2-dichloroethane ($r^2 \leq 0.69$). Because the highest MAC anc MVK mixing ratios occurred downwind of the Syncrude Mildred Lake Facility in samples 3–7, enhancements associated with the oil sands industry cannot be ruled out. However because their strongest correations were with each other, we suggest that their primary local source is likely isoprene oxidation. The average isoprene mixing ratio during Leg 7 was 468±167 pptv. Assuming MAC and MVK formation yields from OH and O3 reactions in the range of 0.16–0.4 (Tani et al., 2010 and references therein), the observed MAC and MVK values during Leg 7 (92±63 and 141±109 pptv, respectively) appear to be within the range that can be explained by isoprene chemistry. Methyl *tert*-butyl ether is a solvent and gasoline additive, and furan is an intermediate in the synthesis of industrial chemicals. The mixing ratio of MTBE remained below its detection limit of 3 pptv throughout Flight 23, showing that it is not being released by the oil sands industry. This is consistent with its declining use in North America because of concerns over groundwater contamination (e.g., Backer, 2010; Lu and Rice, 2010). Like MTBE, furan also remained below its 3 pptv detection limit throughout the flight, in contrast to a maximum mixing ratio of 2344 pptv in very fresh biomass burning plumes that were sampled during ARCTAS.

Halocarbons

Long-lived halocarbons (e.g., CFCs, HCFCs) contribute to stratospheric ozone depletion, and short-lived halocarbons (e.g., $CHCl_3$, C_2Cl_4) can be toxic or carcinogenic. Of the 26 C_1–C_2 halocarbons that we measured, 6 showed both strong enhancements ($\geq 10\%$) over the local background average and larger mixing ratios over the oil sands than at any other time during the Flight, 9 showed minimal enhancements over the local background average, and 11 were not statistically enhanced over the oil sands.

Strongly Enhanced Halocarbons

The first group includes the CFC-substitutes HCFC-22 and HCFC-142b, and the relatively short-lived (2–5 mo) industrial solvents chloroform ($CHCl_3$), trichloroethene (C_2HCl_3), tetrachloroethene (C_2Cl_4) and dichloromethane (CH_2Cl_2). HCFC-22 and HCFC-142b are long-lived refrigerants, foam-blowing agents and solvents whose atmospheric concentrations are rapidly increasing (O'Doherty et al., 2004; Derwent et al., 2007; Montzka et al., 2009). Their mixing ratios over the oil sands were greater than at any other time during the flight, with maximum values of 212 pptv HCFC-22 and 22 pptv HCFC-142b – or respective enhancements of 12% and 18% over the average background (Table 1; Fig. 8a) – showing their usage and release at the mining sites.

The short-lived solvents $CHCl_3$, C_2HCl_3, C_2Cl_4 and CH_2Cl_2 also showed stronger enhancements over the oil sands than during any other portion of the flight (Figs. 4b, 8b). In fact, the maximum C_2HCl_3 and C_2Cl_4 levels (4.8 pptv and 5.9 pptv, respectively; Table 1) were higher in the oil sands plumes than at any other time during the summer phase of ARCTAS. For perspective, these plume values are smaller than average levels measured in urban centers such as Hong Kong, where C_2HCl_3 and C_2Cl_4 mixing ratios of 70±31 and 29±9 pptv were measured during autumn 2007 (Zhang et al., 2010). Maximum respective CH_2Cl_2 and $CHCl_3$ mixing ratios were 35.1 pptv and 15.6 pptv, or 22% and 45% greater than the background average. Dichloromethane, C_2Cl_4, and to a lesser extent $CHCl_3$, were also enhanced relative to background mixing ratios during the descent into Cold Lake, showing their use as solvents in local urban areas as well as by the mining industry.

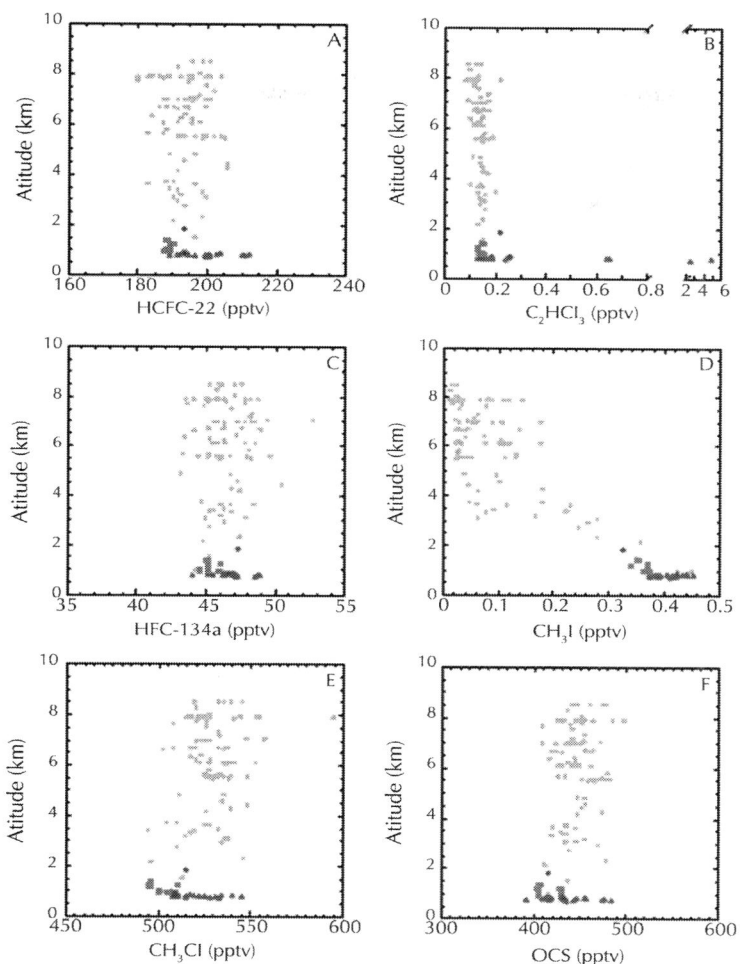

Figure 8: As in Fig. 5 but for selected halocarbons and OCS.

Minimally Enhanced Halocarbons

The second group of compounds, which were minimally enhanced over the oil sands, includes 1,1,1,2-tetrafluoroethane (HFC-134a), some chlorinated species (1,2-dichloroethane, HCFC-141b), the halons (H-1211, H-1301, H-2402) and the methyl halides (CH_3Br, CH_3I, CH_3Cl).

HFC-134a is a CFC-12 replacement that is rapidly increasing in the atmosphere (O'Doherty et al., 2004; Clerbaux and Cunnold, 2007). HFC-134a levels of up to 60 pptv were measured in U.S. pollution plumes (Barletta et al., 2009) compared to a maximum of 48.9 pptv in the oil sands plumes, or 8% greater than the background average, suggesting that the oil sands mining industry is a relatively minor HFC-134a source (Table 1, Fig. 8c). 1,2-Dichloroethane is used as a chemical intermediate and solvent, and its maximum mixing ratio over the oil sands (9.4 pptv) was 16% greater than the background BL average but within the range of mixing ratios measured at other times during the flight, also suggesting relatively small emissions by the oil sands industry (Table 1). Likewise, the CFC replacement compound HCFC-141b showed fairly small (<7%) enhancements over the oil sands (Table 1). The halons are ozone-depleting substances that are still used in North America because of their critical role as fire extinguishing agents (Butler et al., 1998; Fraser et al., 1999; Montzka et al., 2003). Maximum H-1211, H-1301 and H-2402 levels over the oil sands were 5–10% greater than the local background average (Table 1), compared to average H-1211 enhancements of 75% in cities in China (Barletta et al., 2006) where 90% of global H-1211 production occurs (Fraser et al., 1999). Therefore the halon enhancements over the oil sands also appear to be relatively minor, though we note that the H-1301 maximum over the oil sands was greater than at any other time during the summer phase of ARCTAS.

Methyl bromide, an ozone-depleting substance with natural and anthropogenic sources (e.g., oceans, salt marshes, fumigation and biomass burning; Yvon-Lewis et al., 2009 and references therein), had a maximum mixing ratio of 8.3 pptv both over the oil sands and upon final descent into Cold Lake. This 8% enhancement over the local background average was small compared to CH_3Br mixing ratios of up to 15 pptv in fresh biomass burning plumes that were sampled during ARCTAS, suggesting a minor impact of the oil sands industry on CH3Br levels in the sampled plumes.

Methyl iodide is a very short-lived species (1–2 d) with a major oceanic source (Yokouchi et al., 2008) and minor terrestrial sources (e.g., Redeker et al., 2003; Sive et al., 2007). Its mixing ratio declined sharply with altitude, consistent with its short atmospheric lifetime (Fig. 8d). In fact, the shape of the profile and the magnitude of the mixing ratios were remarkably similar to those measured elsewhere over North America during the Intercontinental Chemical Transport Experiment–North America (INTEX-NA) campaign in July and August, 2004 (Sive et al., 2007). In addition, the mixing ratios of CH_3I during Legs 7 and 9 (<0.45 pptv) were at the low end of global background values (0.5–2 pptv; Yokouchi et al., 2008 and references therein) and CH3I did not correlate with other species over the oil sands ($r^2 \leq 0.37$). Therefore the oil sands do not appear to be a significant CH_3I source. The maximum CH_3I mixing ratio was measured in the final sample of Flight Leg 7 (sample 17) and the enhancements of CH_3I during Leg 7 may be attributable to methyl iodide's terrestrial source.

Methyl chloride is the most abundant chlorine-containing compound in the atmosphere. It has a complex budget with multiple sources (e.g., tropical vegetation, biomass burning) and sinks (e.g., OH, soil) (Clerbaux et al., 2007). During the first half of the flight the average CH3Cl mixing ratio in the FT was 530±13 pptv. It then decreased to 503±6 pptv in the background BL, consistent with CH3Cl removal at low altitude by its soil sink. By contrast, the average CH3Cl mixing ratio over the oil sands (522±12 pptv) was enhanced compared to the local background, with a maximum of 545 pptv and a range of values similar to the FT (Fig. 8e). Whereas CH_3Cl showed poor correlation with most compounds over the oil sands (r^2 <0.34), including combustion tracers such as CO and ethyne (r^2 <0.01), it showed some correlation with OCS and HCFC-22 (r^2 = 0.58 for both) followed by HCFC-142b and HFC-134a ($0.46 \leq r^2 \leq 0.47$). Carbonyl sulphide has a strong vegetative sink (Sect. 3.4) and Montzka et al. (2007) found some correlation between OCS and other species with known surface sinks such as CH_3Cl. Therefore the CH_3Cl enhancement is expected to be related to the loss of its soil sink at the mining site, with the possibility of its co-emission from the oil sands industry together with species such as HCFC-22.

Non-enhanced Halocarbons

Carbon tetrachloride (CCl_4), methyl chloroform (CH_3CCl_3) and the chlorofluorocarbons (CFC-11, CFC-12, CFC-113, CFC-114) were not statistically enhanced over the oil sands (Table 1; Fig. 4a). These long-lived compounds have been banned in developed countries since 1996 under the Montreal Protocol, and even though evidence exists for their continued emissions in North America (Millet and Goldstein,

2004; Hurst et al., 2006) their absence from the oil sands mining industry is not surprising. Ethyl chloride ($C2H_5Cl$), another chlorinated hydrocarbon whose industrial demand is declining, also was not enhanced over the oil sands. Likewise, with the exception of methyl bromide (CH_3Br), none of the measured brominated species ($CHBr_3$, CH_2Br_2, $CHBrCl_2$, $CHBr_2Cl$) were enhanced over the oil sands (Table 1). These short-lived species are predominantly emitted from the ocean (e.g., Butler et al., 2007) and their lack of enhancement over the oil sands is not unexpected.

Sulphur Species

Sulphur dioxide (SO_2) is strongly associated with fossil fuel combustion, especially coal and residential oil, and the sulphur content of crude petroleum can be up to 2% (Benkovitz et al., 1996). In the Athabasca oil sands developments, the major SO_2 emission sources are associated with the upgrading and energy production operations at the Suncor and Syncrude sites (Kindzierski and Ranganathan, 2006; D. Spink, personal communication, 2010). For example Suncor produces high sulphur fuel grade petroleum coke from oil sands at its Fort McMurray operations that is 5.7–6.8% sulphur on a dry basis (http://www.suncor.com/en/about/3408.aspx). Some of this coke is used to power its upgrading operations, emitting SO_2 into the atmosphere as a by-product, but most of the coke is buried as a waste product. Whereas Burstyn et al. (2007) measured monthly average SO_2 mixing ratios of 300–1300 pptv in rural locations of western Canada, and Kindzierski and Ranganathan (2006) measured a median outdoor SO_2 mixing ratio of 650 pptv in residential areas of Fort McKay (a small community 64 km north of Fort Mc- Murray, in proximity to the mining operations), SO_2 levels in this study were even smaller during most of the flight, averaging 17±5 pptv in the FT during the first half of the flight and increasing to 102±27 pptv upon descent into the background BL air (Table 1; Fig. 4d). However SO_2 showed remarkable enhancements over the oil sands, with a maximum mixing ratio of 38 730 pptv (38.73 ppbv) in sample 4, or 383× the background BL average. This value is similar to those measured in the urban/industrial environments of heavily polluted cities. For example, Talbot et al. (2008) reported peak airborne SO_2 mixing ratios in excess of 34 000 pptv in BL air sampled over Mexico City during the spring 2006 INTEX-B experiment. Similarly de Foy et al. (2009) measured ground-based SO_2 values exceeding 80 000 pptv in the Mexico City Metropolitan Area during spring 2006, which they attributed to coal combustion, refineries and active volcanoes. Here we attribute the elevated SO_2 levels to coke combustion. The very strong correlations between SO_2 and more than a dozen compounds ($r^2 \geq$ 0.9) – most notably the trimethylbenzenes ($0.98 \leq r^2 \leq 0.99$), NO, NO_2 and NO_y ($0.96 \leq r^2 \leq 0.97$), the ethyltoluenes ($0.90 \leq r^2 \leq 0.97$) and ethene ($r2 = 0.92$) – suggest that coke combustion may be associated with their enhancements as well. Carbonyl sulphide (OCS) is the most abundant sulphurcontaining compound in the remote atmosphere (Ko et al., 2003 and references therein). Its sources include CS_2 oxidation, the oceans, biomass burning, coal burning and aluminum production (e.g., Watts, 2000; Kettle et al., 2002). Like CH_3Cl, the average OCS mixing ratio was fairly constant in the FT during the first half of the flight (445±19 pptv), then diminished with decreasing altitude to 413±13 pptv in background BL air, consistent with its wellknown uptake by vegetation (e.g., Montzka et al., 2007; Blake et al, 2008; Campbell et al., 2008). By contrast OCS was not depleted over the oil sands, where its average mixing ratio (437±26 pptv) was not significantly different than during the first half the flight (Table 1; Fig. 8f). Carbonyl sulphide did not correlate with most compounds over the oil sands ($r^2 \leq 0.38$) but showed some correlation with the same suite of compounds as CH_3Cl, including HCFC-22, HCFC- 142b and HFC-134a ($0.47 \leq r^2 \leq 0.58$). Therefore the relative OCS enhancement over the oil sands compared to the background BL is most likely due to a lack of drawdown from the cleared land in the oil sands area, with the possibility of an OCS source associated with the oil sands industry. Note that even though previous work has shown similarities between OCS and CO_2 because of simultaneous uptake by photosynthetically active vegetation during the daytime (Montzka et al., 2007; Campbell et al., 2008; White et al., 2010), here OCS and CO_2 showed poor correlation ($r^2 = 0.16$) because CO_2 has a clear additional fossil fuel source at the oil sands developments (Sect. 3.6). Dimethyl sulphide (DMS) is a short-lived sulphur species (1–2 days) with a major oceanic source and minor sources including vegetation and biomass burning (e.g., Watts et al., 2000; Gondwe et al., 2003; Meinardi et al., 2003). Like CH3I, DMS levels strongly increased with decreasing altitude. Dimethyl sulphide levels were below detection (<1 pptv) in the FT, increasing to 4.7±0.8 pptv in the background BL. Its average mixing ratio doubled to

10.7±3.9 pptv over the oil sands with a maximum enhancement of 18 pptv (Table 1). For comparison, DMS levels over productive oceanic regions are on the order of 100–250 pptv (e.g., Nowak et al., 2001). Dimethyl sulphide was most strongly enhanced in samples 4, 5 and 6 and correlated with a range of compounds, most strongly CO_2 (r^2 =0.82) as well as H-1211, 1,2-dichloroethane and HCFC-142b ($0.71 \le r^2 \le 0.74$), indicating that its oil sands source is industrial.

NO, NO_2, NO_y and O_3

Like SO_2, the major source of NO_x is fossil fuel combustion (Benkovitz et al., 1996). Nitric oxide, NO_2 and NO_y were all very strongly enhanced over the oil sands compared to their respective background values of 16±6 pptv, 24±11 pptv and 194±33 pptv (Table 1). The maximum NO, NO_2 and NO_y mixing ratios occurred in sample 4 and were 4980, 4995 and 10 555 pptv, respectively, representing enhancements of 319×, 210× and 54× the local background (Table 1; Fig. 9a–b). Recall that these numbers are based on the average of those NO, NO_2 and NO_y measurements that overlapped the VOC sampling times (Sect. 3). The maximum NO, NO_2 and NO_y values over the oil sands based on 1 s measurements were even higher: 9545, 9205 and 21 800 pptv, respectively. The NO maximum was the highest recorded throughout the summer phase of ARCTAS, showing the very strong emissions of nitrogen oxides from the mining industry. The NO_y levels in the oil sands plumes lie within the lower range of values measured in megacities such as Tokyo, Mexico City and Beijing, which can vary from 2000–200 000 pptv (Parrish et al., 2009). The major NO_x sources at the mining sites are (1) the upgraders, gaseous fuel fired boilers, heaters, and co-generation units for heat and power production, and (2) the heavy hauler mine fleets (D. Spink, pers. comm., 2010). Because NO, NO_2 and NO_y correlated perfectly with each other ($r^2 = 1.00$) and very strongly with SO_2, most C_2–C_4 alkenes, and the C_9 aromatics ($0.90 \le r^2 \le 0.99$) (Table 3), we conclude that the observed emissions were from the upgraders, etc. rather than the heavy hauler fleet.

Ozone was not enhanced over the oil sands, and its maximum mixing ratio (31 ppbv) was the same as the average background BL value (31±1 ppbv) (Table 1, Fig. 9c). In fact O_3 was anti-correlated with NO over the oil sands (Fig. 10) because the very strong NO emissions led to a titration of O_3, a loss process which dominated over

O_3 production on this short time-scale. Note that in Fig. 10 virtually all (_95%) of the NO_y is in the form of NO_x.

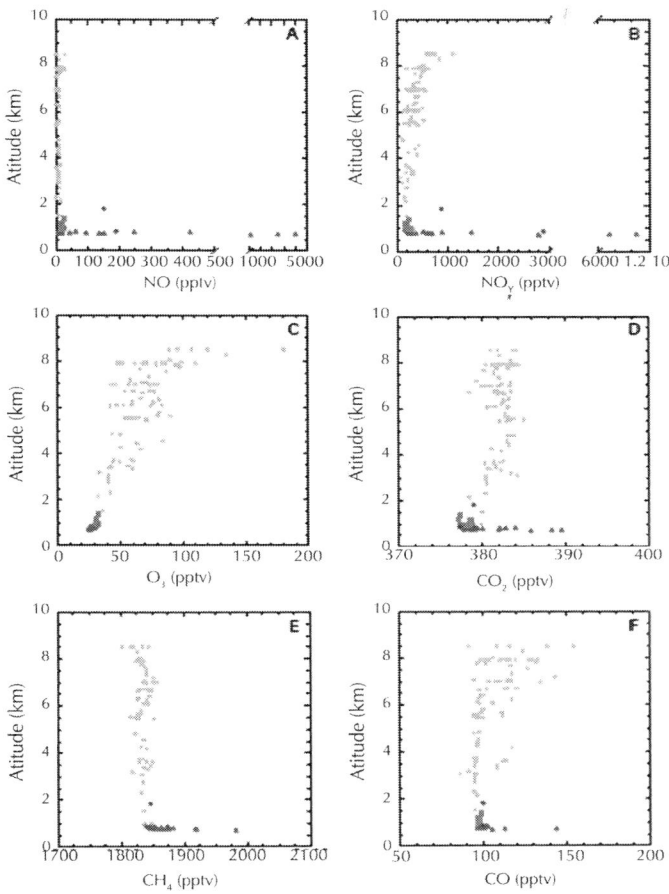

Figure 9: As in Fig. 5 but for (a) NO, (b) NO_y, (c) O_3, (d) CO_2, (e) CH_4 and (f) CO.

CO_2, CH_4 and CO

Carbon dioxide (CO_2) is the leading contributor to the enhanced radiative forcing of the atmosphere, followed by methane (CH_4) (Forster et al., 2007). Compared to the FT, CO_2 was relatively depleted in the background BL air during Flight 23 (Table 1, Fig. 9d), consistent with its summertime uptake by terrestrial vegetation (Erickson et al., 1996; Randerson et al., 1997, 1999). By contrast CO_2 showed a clear enhancement over the oil sands, with a maximum mixing ratio of 389 ppmv that is outside the range of values measured during the rest of the flight. 114C in CO_2 was depleted in these enhanced samples, indicating a fossil fuel CO_2 influence (not shown). Carbon dioxide correlated with a wide range of compounds associated with the mining industry, including DMS, combustion tracers such as i-butene, ethene, ethyne and NO_y, and industrial tracers such as MEK, 2,3-dimethylbutane, $CHCl_3$ and 3-ethyltoluene ($0.72 \leq r^2 \leq 0.82$), showing its emissions during many stages of the mining operations.

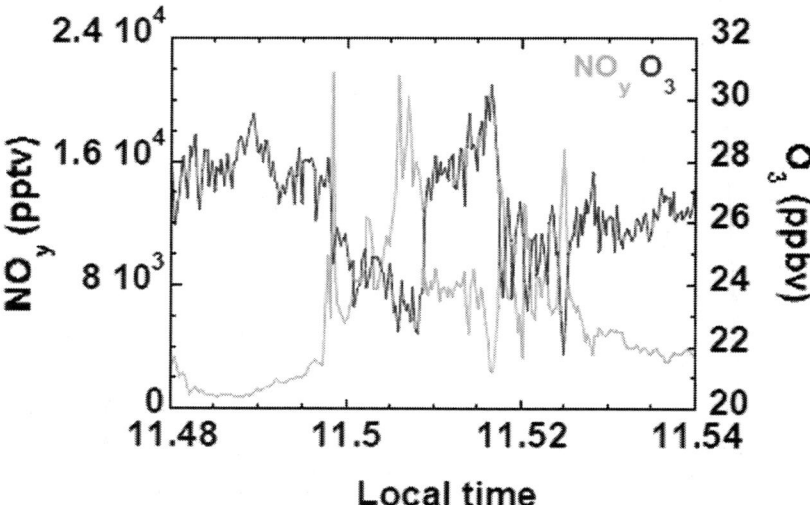

Figure 10: High-resolution time series of NO_y (green) and O_3 (blue) over the oil sands during the first half of Flight Leg 7, showing an anti-correlation between O_3 and NO_y in the plumes.

Methane is long-lived and well-mixed in the remote atmosphere, and its sources include biogenic sources (e.g., wetlands, ruminants, rice agriculture), biomass burning, and fossil fuel mining and burning (Denman et al., 2007). Its mixing ratio varied little in the FT during Flight 23, averaging 1836±10 ppbv (Table 1; Fig. 9e). Unfortunately

CH_4 measurements were not available during the background BL run (Leg 9). However CH_4 shows a characteristic north-south gradient (e.g., Simpson et al., 2002), and because Cold Lake is at a similar latitude to Leg 9 and does not appear to have been appreciably impacted by local CH_4 sources (Fig. 9e), the average CH_4 mixing ratio in the BL during the final descent into Cold Lake (1843±5 ppbv) is expected to be a reasonable proxy for the local background at this time of year. Methane mixing ratios were clearly enhanced over the oil sands, reaching 1983 ppbv in sample 6. By comparison the maximum CH_4 mixing ratio during the summer phase of ARCTAS (2000 ppbv) was measured in a fresh biomass burning plume. Natural gas is heavily used by the oil sands mining industry (Sect. 1), but the CH4 results are surprising in that CH_4 did not correlate particularly well with ethane and propane (0.52 ≤ r^2 ≤ 0.57), which, like CH_4, are components of natural gas. The composition of commercialgrade natural gas is variable, ranging from 70–95% CH_4, with the remainder primarily ethane (e.g., Xiao et al., 2008; McTaggart-Cowan et al., 2010). Although we are not aware of the composition of the natural gas that is typically used by the oil sands industry, a specs sheet for natural gas from western Canada suggests the following typical composition on a per mole percentage basis: 94.6% CH_4, 2.5% ethane, 0.2% propane, 0.03% n-butane, etc. (http://www.naesb.org/pdf2/ wgq bps100605w2.pdf). This suggests a relative release rate of 38 moles of CH_4 to 1 mole of ethane, which is more ethane than we observed. The maximum CH_4 mixing ratio over the oil sands (sample 6) represented an excess of 140 ppbv over its local background value of 1843 ppbv. The ethane mixing ratio in sample 6 was 1.106 ppbv, representing an excess of 0.325 ppbv. If the CH_4 enhancement were entirely due to natural gas emissions, we would have expected to measure a corresponding ethane excess of 3.7 ppbv. Therefore the measured ethane enhancement was about 10× smaller than would be expected from natural gas emissions, which suggests low natural gas leakage levels from the mining operations. Similarly, hydrocracking operations do not appear likely to explain the observed CH_4 enhancements. The catalytic hydrocracking of Athabasca bitumen vacuum bottoms has been found to give a weight per cent yield of 9.3% CH_4, 10.5% ethane, 15.0% propane and 9.6% C_4H_{10} (Dehkissia et al., 2004), or relative molar yields of roughly 4:2:2:1 for CH_4:ethane:propane:butanes. Again, this represents much more ethane than we observed, and assuming these hydrocracking yields apply here and are equivalent to emission rates, it does not appear that hydrocracking can be responsible for the high CH_4 excesses that were observed. Instead we suggest that the CH_4 enhancements are likely related to tailings pond emissions. As stated in Sect. 1, the Syncrude Mildred Lake tailings pond became methanogenic in the 1990s and releases up to 10 gCH_4 m^{-2} d^{-1}. Ethane is not produced in anaerobic environments, and tailings pond emissions of CH_4 explain the relatively large CH_4 enhancements compared to ethane as well as the relatively poor correlation between CH_4 and ethane.

Carbon monoxide (CO) is a potentially toxic gas that is also a precursor to photochemical smog. The maximum CO mixing ratio over the oil sands (144 ppbv in sample 5, or 48% greater than the local background of 97±1 ppbv) was comparable to mixing ratios measured at other times during the flight (Table 1; Fig. 9f) and is not considered a large enhancement. For example Baker et al. (2008) found typical summertime CO mixing ratios of 300±66 ppbv in US cities, and CO mixing ratios measured near fresh biomass burning plumes during ARCTAS exceeded 1900 ppbv (not shown). Interestingly, CO correlated most strongly with the suite of C_4–C_9 alkanes, C_6–C_8 aromatics and C_5–C_6 cycloalkanes (0.85 ≤ r^2 ≤ 0.97) that were associated with direct emissions from the oil sands and/or diluent (Sect. 3.2). It also showed good correlations with $CHCl_3$ (r^2 =0.84) and the C_9 aromatics (0.66<r^2 <0.83). Carbon monoxide is an urban/industrial combustion tracer and these results suggest that relatively low levels of CO are emitted throughout the mining operations.

CONCLUSIONS

Mixing ratios of 84 trace gases were measured in boundary layer air over oil sands surface mining operations in northern Alberta on 10 July 2008 aboard the NASA DC-8 research aircraft as part of the ARCTAS mission. Compared to local background air, 15 of these compounds showed no statistical enhancements over the oil sands (propyne, MTBE, furan, CFC-11, CFC-12, CFC-113, CFC-114, CCl4, CH3CCl3, ethyl chloride, $CHBr_3$, CH_2Br_2, $CHBrCl_2$, $CHBr_2Cl$ and O_3). As is to be expected with high NO levels, O_3 was anticorrelated with NO and it appears that depletion of O_3 by NO dominated over O_3 production on this short time-scale. Another 9 compounds showed minimal enhancements over the oil sands (HFC-134a, HCFC-141b, 1,2-dichloroethane, H-1211, H-1301, H-2402, CH_3Br, CH_3I, CH_3Cl).

Carbon monoxide showed a clear enhancement (up to 48%) and strong correlations with many other compounds, but its maximum mixing ratio of the flight was not over the oil sands. By contrast the remaining 59 compounds showed greater mixing ratios over the oil sands than at any other time during the flight, and in some cases than at any other time during the entire mission. The maximum enhancements were 1.1–397× the background BL values, most notably nheptane (397×), SO_2 (382×), NO (319×), n-octane (242×) and methylcyclohexane (113×). Based on their mutual correlations, the elevated trace gases fell into two groups. The first group includes CO and species associated with direct evaporative emissions from the oil sands themselves and/or from the diluent used to lower the viscosity of the recovered bitumen (C_4–C_9 alkanes, C_5–C_6 cycloalkanes, C_6–C_8 aromatics). By contrast to the n-alkanes, the enhancements of aromatics such as benzene were relatively small, especially compared to values that have been measured downstream at petrochemical refineries. The second group also includes CO in addition to a wide variety of species associated with emissions from the mining effort, for example coke combustion (e.g., SO_2) and upgrading (e.g., NO, NO_2, NO_y). The maximum SO_2 and NO levels over the oil sands were greater than at any other time during the summer phase of ARCTAS, and even though northern Alberta is a rural environment, the SO_2 and NO_y levels were comparable to those measured in the world's megacities. The CO enhancements were generally small suggesting that the upgraders emit low levels of CO, which is to be expected for high temperature combustion where the carbon is converted mostly to CO_2 (i.e., efficient combustion). Although the oil sands industry is a major user of natural gas, a strong natural gas signal was not evident in the data suggesting low natural gas leakage levels and high efficiency combustion associated with the upgraders. Instead, the elevated CH_4 levels are consistent with methanogenic tailings pond emissions. Like CO, the butanes also fell into both groups and appear to have multiple sources from oil sands mining.

Isoprene, a biogenic tracer, was enhanced during the oil sands boundary layer run but with a maximum over a vegetated area south of the major mining operations. By contrast, the monoterpenes _-pinene and _-pinene (also biogenic tracers) were most enhanced in the oil sands plumes and showed good correlations with many industrial species associated with the mining effort including C_2Cl_4 and SO_2. The possibility of pinene emissions directly associated with the mining operations requires further study. Carbonyl sulphide and CH_3Cl were also notable in that they failed to be drawn down over the surface mining sites, most likely because of the removal of their vegetation and soil sinks, respectively.

These measurements represent the only independent characterization of trace gas emissions from oil sands mining operations of which we are aware. Although the absolute mixing ratios of many VOCs were relatively modest compared to major petrochemical facilities that have been studied, they are significantly enhanced above background levels. The high reactivity of most of these gases, combined with significant emissions of NO_x and SO_2 in what would otherwise be a relatively pristine area, mean that they do have the potential to form O_3 and acid conditions downwind of this activity. Further study of such potential effects is required. For example modeling and a multi-day day ground-based grid study near the mining sites would help to more completely characterize the trace gases that are emitted from and impacted by the Alberta oil sands industry.

ACKNOWLEDGEMENTS

We thank the ARCTAS crew and science team for their hard work throughout the mission, and we gratefully acknowledge helpful discussions with many of our colleagues, especially Jim Crawford (NASA Langley) and Joost de Gouw (NOAA/ESRL). We also thank David Spink (Fort McKay IRC) for many helpful comments on the manuscript. This research was funded by NASA grant NNX09AB22G.

REFERENCES

1.   Alboudwarej, H., Felix, J., Taylor, S., Badry, R., Bremner, C., Brough, B., Skeates, C., Baker, A., Palmer, D., Pattison, K., Beshry, M., Krawchuk, P., Brown, G., Calvo, R., Ca~nas Triana, J. A., Hathcock, R., Koerner, K. Hughes, T., Kundu, D., de C´ardenas, J. L., andWest, C.: Highlighting heavy oil, Oilfield Rev., 34–53, 2006.

2. Apel, E. C., Calvert, J. G., Gilpin, T. M., Fehsenfeld, F. C., Parrish, D. D., and Lonneman, W. A.: The Nonmethane Hydrocarbon Intercomparison Experiment (NOMHICE): Task 3, J. Geophys. Res., 104(21), 26069–26086, 1999.

3. Apel, E. C., Calvert, J. G., Gilpin, T. M., Fehsenfeld, F., and Lonneman, W. A.: Nonmethane Hydrocarbon Intercomparison Experiment (NOMHICE): Task 4, ambient air, J. Geophys. Res., 108(D9), 4300, doi: 10.1029/2002JD002936, 2003.

4. Atkinson, R.: Gas-phase tropospheric chemistry of organic compounds, J. Phys. Chem. Ref. Data, Monograph, 2, 1–216, 1994.

5. Atkinson, R.: Atmospheric chemistry of VOCs and NOx, Atmos. Environ., 34, 2063–2101, 2000.

6. Atkinson, R. and Arey, J.: Atmospheric degradation of volatile organic compounds, Chem. Rev., 103, 4605–4638, 2003.

7. Atkinson, R., Arey, J., Hoover, S., and Preston, K.: Atmospheric chemistry of gasoline-related emissions: Formulation of pollutants of potential concern, Draft for public comment, California Environmental Protection Agency, 2005.

8. Backer, W. S.: Contamination of drinking water by methyl tertiarybutyl ether (MTBE) and its effect on plasma enzymes, Sci. Res. Essays, 5(14), 1809–1812, 2010.

9. Baker, A. K., Beyersdorf, A. J., Doezema, L. A., Katzenstein, A., Meinardi, S., Simpson, I. J., Blake, D. R., and Rowland, F. S.: Measurements of nonmethane hydrocarbons in 28 United States cities, Atmos. Environ., 42, 170–182, 2008.

10. Barletta, B., Meinardi, S., Simpson, I. J., Khwaja, H. A., Blake, D. R., and Rowland, F. S.: Mixing ratios of volatile organic compounds (VOCs) in the atmosphere of Karachi, Pakistan, Atmos. Environ., 36, 3429–3443, 2002

11. Barletta, B., Meinardi, S., Rowland, F. S., Chan, C.-Y., Wang, X., Zou, S., Chan, L. Y., and Blake, D. R.: Volatile organic compounds in 43 Chinese cities, Atmos. Environ., 39, 5979–5990, 2005.

12. Barletta, B., Meinardi, S., Simpson, I. J., Atlas, E. L., Beyersdorf, A. J., Baker, A. K., Blake, N. J., Yang, M., Midyett, J. R., Novak, B. J., McKeachie, R. J., Fuelberg, H. E., Sachse, G. W., Avery, M. A., Campos, T., Weinheimer, A. J., Rowland, F. S., and Blake, D. R.: Characterization of volatile organic compounds (VOCs)

in Asian and north American pollution plumes during INTEX-B: identification of specific Chinese air mass tracers, Atmos. Chem. Phys., 9, 5371–5388, doi:10.5194/acp-9-5371-2009, 2009.

13. Benkovitz, C. M., Scholtz, M. T., Pacyna, J., Tarras´on, L., Dignon, J., Voldner, E. C., Spiro, P. A., Logan J. A., and Graedel, T. E.: Global gridded inventories of anthropogenic emissions of sulphur and nitrogen, J. Geophys. Res., 101(D22), 29239–29253, 1996.

14. Blake, D. R. and Rowland, F. S.: Continuing world-wide increase in tropospheric methane, 1978–1987, Science, 239, 1129–1131, 1988.

15. Blake, D. R. and Rowland, F. S.: Urban leakage of liquefied petroleum gas and its impacts on Mexico City air quality, Science, 269, 953–956, 1995.

16. Blake, N. J., Blake, D. R., Sive, B. C., Chen, T.-Y., Rowland, F. S., Collins Jr., J. E., Sachse, G. W., and Anderson, B. E.: Biomass burning emissions and vertical distribution of atmospheric methyl halides and other reduced carbon gases in the South Atlantic region, J. Geophys. Res., 101(D19), 24151– 24164, 1996.

17. Blake, N. J., Blake, D. R., Simpson, I. J., Meinardi, S., Swanson, A. L., Lopez, J. P., Katzenstein, A. S., Barletta, B., Shirai, T., Atlas, E., Sachse, G., Avery, M., Vay, S., Fuelberg, H. E., Kiley, C. M., Kita, K., and Rowland, F. S.: NMHCs and halocarbons in Asian continental outflow during the Transport and Chemical Evolution over the Pacific (TRACE-P) field campaign: Comparison with PEM-West B, J. Geophys. Res., 108(D20), 8806, doi: 10.1029/2002JD003367, 2003.

18. Blake, N. J., Campbell, J. E., Vay, S. A., Fuelberg, H. E., Huey, L. G., Sachse, G., Meinardi, S., Beyersdorf, A., Baker, A., Barletta, B., Midyett, J., Doezema, L., Kamboures, M., McAdams, J., Novak, B., Rowland, F. S., and Blake, D. R.: Carbonyl sulfide (OCS): Large-scale distributions over North America during INTEX-NA and relationship to CO2, J. Geophys. Res., 113, D09S90, doi: 10.1029/2007JD009163, 2008.

19. Borbon, A., Fontaine, H., Veillerot, M., Locoge, N., Galloo, J. C., and Guillermo, R.: An investigation into the traffic-related fraction of isoprene at an urban location, Atmos. Environ., 35, 3749– 3760, 2001.

20. Burstyn, I., Senthilselvan, A., Kim, H.-M., Cherry, N. M., Pietroniro, E., and Waldner, C.: Industrial sources influence air concentrations of hydrogen sulfide and sulfur dioxide in rural areas of western Canada, J. Air Waste Manage., 57(10), 1241–1250, 2007.

21. Butler, J. H., Montzka, S. A., Clarke, A. D., Lobert, J. M., and Elkins, J. W.: Growth and distribution of halons in the atmosphere, J. Geophys. Res., 103(D1), 1503–1511, 1998.

22. Butler, J. H., King, D. B., and Lobert, J. M.: Oceanic distributions and emissions of short-lived halocarbons, Global Biogeochem. CY. 21, GB1023, doi: 10.1029/2006GB002732, 2007.

23. Buzcu, B. and Fraser, M. P.: Source identification and apportionment of volatile organic compounds in Houston, TX, Atmos. Environ., 40, 2385–2400, 2006.

24. Campbell, J. E., Carmichael, G. R., Chai, T., Mena-Carrasco, M., Tang, Y., Blake, D. R., Blake, N. J., Vay, S. A., Collatz, G. J., Baker, I., Berry, J. A., Montzka, S. A., Sweeney, C., Schnoor, J. L., and Stanier, C. O.: Photosynthetic control of atmospheric carbonyl sulfide during the growing season, Science, 322, 1085–1088, 2008.

25. Cetin, E., Odabasi, M., and Seyfioglu, R.: Ambient volatile organic compound (VOC) concentrations around a petrochemical complex and a petroleum refinery, Sci. Total Environ., 312(1–30), 103–112, 2003.

26. Choi, Y., Elliott, S., Simpson, I. J., Blake, D. R., Colman, J. J., Dubey, M. K., Meinardi, S., Rowland, F. S., Shirai, T., and Smith, F. A.: Survey of whole air data from the second airborne biomass burning and lightning experiment using principal component analysis, J. Geophys. Res., 108(D5), 4163, doi: 10.1029/2002JD002841, 2003.

27. Colman, J. J., Swanson, A. L., Meinardi, S., Sive, B. C., Blake, D. R., and Rowland, F. S.: Description of the analysis of a wide range of volatile organic compounds in whole air samples collected during PEM-Tropics A and B, Anal. Chem., 73, 3723–3731, 2001.

28. Clerbaux, C., Cunnold, D. M., Anderson, J., Engel, A., Fraser, P. F., Mahieu, E., Manning, A., Miller, J., Montzka, S. A., Nassar, R., Prinn, R., Reimann, S., Rinsland, C. P., Simmonds, P., Verdonik,

D., Weiss, R., Wuebbles, D., and Yokouchi, Y.: Long-lived compounds, Scientific Assessment of Ozone Depletion: 2006 Global Ozone Research and Monitoring Project, Report no. 50, Chapter 1, World Meteorological Organization, Geneva, 2007.

29. de Foy, B., Krotkov, N. A., Bei, N., Herndon, S. C., Huey, L. G., Mart´ınez, A.-P., Ruiz-Su´arez, L. G., Wood, E. C., Zavala, M., and Molina, L. T.: Hit from both sides: tracking industrial and volcanic plumes in Mexico City with surface measurements and OMI SO2 retrievals during the MILAGRO field campaign, Atmos. Chem. Phys., 9, 9599–9617, doi:10.5194/acp-9-9599-2009, 2009.

30. de Gouw, J. A., Te Lintel Hekkert, S., Mellqvist, J., Warneke, C., Atlas, E. L., Fehsenfeld, F. C. Fried, A., Frost, G. J., Harren, F. J. M., Holloway, J. S., Lefer, B., Lueb, R., Meagher, J. F., Parrish, D. D., Patel, M., Pope, L., Richter, D., Rivera, C., Ryerson, T. B., Samuelsson, J., Walega, J., Washenfelder, R. A., Weibring, P., and Zhu, X.: Airborne measurements of ethene from industrial sources using Laser Photo-Acoustic Spectroscopy, Environ. Sci. Technol., 43(7), 2437–2442, 2009.

31. Dehkissia, S., Larachi, F., and Chornet, E.: Catalytic (Mo) upgrading of Athabasca bitumen vacuum bottoms via two-step hydrocracking and enhancement of Mo-heavy oil interaction, Fuel, 83, 1323–1331, 2004.

32. Denman, K. L., Brasseur, G., Chidthaisong, A., Ciais, P., Cox, P. M., Dickinson, R. E., Hauglustaine, D., Heinze, C., Holland, E., Jacob, D., Lohmann, U., Ramachandran, S., da Silva Dias, P. L., Wofsy, S. C., and Zhang, X.: Couplings between changes in the climate system and biogeochemistry, in: Climate Change 2007: The Physical Science Basis. Contribution of Working Group I to the Fourth Assessment Report of the Intergovernmental Panel on Climate Change, edited by: Solomon, S., Qin, D., Manning, M., Chen, Z., Marquis, M., Averyt, K. B., Tignor, M., and Miller, H. L., Cambridge University Press, Cambridge, UK and New York, NY, USA, 2007.

33. Derwent, R. G., Simmonds, P. G., Greally, B. R., O'Doherty, S., McCulloch, A., Manning, A., Reimann, S., Folini, D., and Vollmer, M. K.: The phase-in and phase-out of European emissions of HCFC-141b and HCFC-142b under the Montreal

Protocol: Evidence from observations at Mace Head, Ireland and Jungfraujoch, Switzerland from 1994 to 2004, Atmos. Environ., 41, 757–767, 2007.

34. Ehhalt, D. H. and Prather, M.: Atmospheric chemistry and greenhouse gases, in Climate Change 2001: The Scientific Basis, 245–287, Cambridge Univ. Press, New York, 2001.

35. Erickson, D. J., Rasch, P. J., Tans, P. P., Friedlingstein, P., Ciais, P., Maier-Reimer, E., Six, K., Fischer, C. A., and Walters, S.: The seasonal cycle of atmospheric $CO2$: A study based on the NCAR Community Climate Model (CCM2), J. Geophys. Res., 101(D10), 15079–15097, 1996.

36. Folberth, G. A., Hauglustaine, D. A., Lathi`ere, J., and Brocheton, F.: Interactive chemistry in the Laboratoire de Mtorologie Dynamique general circulation model: model description and impact analysis of biogenic hydrocarbons on tropospheric chemistry, Atmos. Chem. Phys., 6, 2273–2319, doi:10.5194/acp-6- 2273-2006, 2006.

37. Forster, P., Ramaswamy, V., Artaxo, P., Berntsen, T., Betts, R., Fahey, D. W., Haywood, J., Lean, J., Lowe, D. C., Myhre, G., Nganga, J., Prinn, R., Raga, G., Schulz, M., and van Dorland, R.: Changes in Atmospheric Constituents and in Radiative Forcing, in: Climate Change 2007: The Physical Science Basis. Contribution of Working Group I to the Fourth Assessment Report of the Intergovernmental Panel on Climate Change, edited by: Solomon, S., Qin, D., Manning, M., Chen, Z., Marquis, M., Averyt, K. B., Tignor, M., and Miller, H. L., Cambridge University Press, Cambridge, UK and New York, NY, USA, 2007.

38. Fraser, P. J., Oram, D. E., Reeves, C. E., Penkett, S. A., and Mc-Culloch, A.: Southern hemispheric halon trends (1978–1998) and global halon emissions, J. Geophys. Res., 104(D13), 15985–16000, 1999.

39. Fried, A., Diskin, G., Weibring, P., Richter, D., Walega, J. G., Sachse, G., Slate, T., Rana, M., and Podolske, J.: Tunable infrared laser instruments for airborne atmospheric studies, Appl. Phys. B, 92(3), 409–417, 2008.

40. Fuelberg, H. E., Harrigan, D. L., and Sessions, W.: A meteorological overview of the ARCTAS 2008 mission, Atmos. Chem. Phys., 10, 817–842, doi:10.5194/acp-10-817-2010, 2010.

41. Fuentes, J. D., Lerdau, M., Atkinson, R., Baldocchi, D., Bottenheim, J. W., Ciccioli, P., Lamb, B., Geron, C., Gu, L., Guenther, A., Sharkey, T. D., and Stockwell, W.: Biogenic hydrocarbons in the atmospheric boundary layer: A review, B. Am. Meteorol Soc., 81(7), 1537–1575, 2000.

42. Gilman, J. B., Kuster, W. C., Goldan, P. D., Herndon, S. C., Zahniser, M. S., Tucker, S. C., Brewer, W. A., Lerner, B. M., Williams, E. J., Harley, R. A., Fehsenfeld, F. C., Warneke, C., and de Gouw, J. A.: Measurements of volatile organic compounds during the 2006 TexAQS/GoMACCS campaign: Industrial influences, regional characteristics, and diurnal dependencies of the OH reactivity, J. Geophys. Res., 114, D00F06, doi:10.1029/2008JD011525, 2009.

43. Gondwe, M., Krol, M., Gieskes, W., Klaassen, W., and de Baar, H.: The contribution of ocean-leaving DMS to the global atmospheric burdens of DMS, MSA, SO2, and NSS SO=4 , Global Biogeochem. Cy., 17(2), 1056, doi:10.1029/2002GB001937, 2003.

44. Guenther, A., Geron, C., Pierce, T., Lamb, B., Harley, P., and Fall, R.: Natural emissions of non-methane volatile organic compounds, carbon monoxide, and oxides of nitrogen from North America, Atmos. Environ., 34, 2205–2230, 2000.

45. Guenther, A., Karl, T., Harley, P., Wiedinmyer, C., Palmer, P. I., and Geron, C.: Estimates of global terrestrial isoprene emissions using MEGAN (Model of Emissions of Gases and Aerosols from Nature), Atmos. Chem. Phys., 6, 3181–3210, doi:10.5194/acp-6- 3181-2006, 2006.

46. Holopainen, J. K. and Gershenzon, J.: Multiple stress factors and the emission of plant VOCs, Trends Plant Sci., 15(3), 176–184, 2010.

47. Holowenko, F. M., MacKinnon, M. D., and Fedorak, P. M.: Methanogens and sulfate-reducing bacteria in oil sands fine tailings waste, Can. J. Microbiol., 46(10), 927–937, 2000.

48. Horowitz, L. W., Walters, S., Mauzerall, D. L., Emmons, L. K., Rasch, P. J., Granier, C., Tie, X., Lamarque, J.-F., Schultz, M. G., Tyndall, G. S., Orlando, J. J., and Brasseur, G. P.: A global simulation of tropospheric ozone and related tracers: Description

and evaluation of MOZART, version 2, J. Geophys. Res., 108(D24), 4784, doi:10.1029/2002JD002853, 2003.

49. Hurst, D. F., Lin, J. C., Romashkin, P. A., Daube, B. C., Gerbig, C., Matross, D. M.,Wofsy, S. C., Hall, B. D., and Elkins, J.W.: Continuing global significance of emissions of Montreal Protocolrestricted halocarbons in the USA and Canada, J. Geophys. Res., 111, D15302, doi:10.1029/2005JD006785, 2006.

50. Jacob, D. J., Field, B. D., Jin, E. M., Bey, I., Li, Q., Logan, J. A., Yantosca, R. M., and Singh, H. B.: Atmospheric budget of acetone, J. Geophys. Res., 107(D10), 4100, doi:10.1029/2001JD000694, 2002.

51. Jacob, D. J., Field, B. D., Li, Q., Blake, D. R., de Gouw, J., Warneke, C., Hansel, A., Wisthaler, A., Singh, H. B., and Guenther, A.: Global budget of methanol: Constraints from atmospheric observations, J. Geophys. Res., 110, D08303, doi:10.1029/2004JD005172, 2005.

52. Jacob, D. J., Crawford, J. H., Maring, H., Clarke, A. D., Dibb, J. E., Emmons, L. K., Ferrare, R. A., Hostetler, C. A., Russell, P. B., Singh, H. B., Thompson, A. M., Shaw, G. E., McCauley, E., Pederson, J. R., and Fisher, J. A.: The Arctic Research of the Composition of the Troposphere from Aircraft and Satellites (ARCTAS) mission: design, execution, and first results, Atmos. Chem. Phys., 10, 5191–5212, doi:10.5194/acp-10-5191-2010, 2010.

53. Jordan, C., Fitz, E., Hagan, T., Sive, B., Frinak, E., Haase, K., Cottrell, L., Buckley, S., and Talbot, R.: Long-term study of VOCs measured with PTR-MS at a rural site in New Hampshire with urban influences, Atmos. Chem. Phys., 9, 4677–4697, doi:10.5194/acp-9-4677-2009, 2009.

54. Kabir, E. and Kim, K.-H.: An on-line analysis of 7 odorous volatile organic compounds in the ambient air surrounding a large industrial complex, Atmos. Environ., 44, 3492–3502, 2010.

55. Karl, T., Apel, E., Hodzic, A., Riemer, D. D., Blake, D. R., and Wiedinmyer, C.: Emissions of volatile organic compounds inferred from airborne flux measurements over a megacity, Atmos. Chem. Phys., 9, 271–285, doi:10.5194/acp-9-271-2009, 2009a.

56. Karl, T., Guenther, A., Turnipseed, A., Tyndall, G., Artaxo, P., and Martin, S.: Rapid formation of isoprene photo-oxidation products

observed in Amazonia, Atmos. Chem. Phys., 9, 7753–7767, doi:10.5194/acp-9-7753-2009, 2009b.

57.   Katzenstein, A. S., Doezema, L. A., Simpson, I. J., Blake, D. R., and Rowland, F. S.: Extensive regional atmospheric hydrocarbon pollution in the southwestern United States, P. Natl. Acad. Sci., 100, 11975–11979, 2003.

58.   Kesselmeier, J. and Staudt, M., Biogenic volatile organic compounds (VOC): An overview on emission, physiology and ecology, J. Atmos. Chem., 33, 23–88, 1999.

59.   Kettle, A. J., Kuhn, U., von Hobe, M., Kesselmeier, J., and Andreae, M. O.: Global budget of atmospheric carbonyl sulfide: Temporal and spatial variations of the dominant sources and sinks, J. Geophys. Res., 107(D22), 4658, doi:10.1029/2002JD002187, 2002.

60.   Kim, S., Huey, L. G., Stickel, R. E., Tanner, D. J., Crawford, J. H., Olson, J. R., Chen, G., Brune, W. H., Ren, X., Lesher, R., Wooldridge, P. J., Bertram, T. H., Perring, A., Cohen, R. C., Lefer, B. L., Shetter, R. E., Avery, M., Diskin, G., and Sokolik, I.: Measurement of HO2NO2 in the free troposphere during the Intercontinental Chemical Transport Experiment-North America 2004, J. Geophys. Res., 112, D12S01, doi:10.1029/2006JD007676, 2007.

61.   Kindzierski, W. B. and Ranganathan, H. K. S.: Indoor and outdoor SO2 in a community near oil sand extraction and production facilities in northern Alberta, J. Environ. Eng. Sci., 5, S121–S129, 2006.

62.   Ko, M. K. W., Poulet, G., Blake, D. R., Boucher, O., Burkholder, J. H., Chin, M., Cox, R. A., George, C., Graf, H.-F., Holton, J. R., Jacob, D. J., Law, J. S., Lawrence, M. G., Midgley, P. M., Seakins, P. W., Shallcross, D. E., Strahan, S. E., Wuebbles, D. J., and Yokouchi, Y.: Very short-lived halogen and sulfur substances, Scientific Assessment of Ozone Depletion: 2002 Global Ozone Research and Monitoring Project, Report no. 47, Chapter 2, World Meteorological Organization, Geneva, 2003.

63.   Lu, N. and Rice, R. W.: Demand drivers and price supports for bioethanol use as fuel in the United States: A brief review, For. Prod. J., 60(2), 126–132, 2010.

64.   McTaggart-Cowan, G. P., Rogak, S. N., Munshi, S. R., Hill, P. G., and Bushe, W. K.: The influence of fuel composition on a

heavy-duty, natural-gas direct-injection engine, Fuel, 89, 752–759, 2010.

65. Meinardi, S., Simpson, I. J., Blake, N. J., Blake, D. R., and Rowland, F. S.: Dimethyl disulfide (DMDS) and dimethyl sulfide (DMS) emissions from biomass burning in Australia, Geophys. Res. Lett., 30(9), 1454, doi:10.1029/2003GL016967, 2003.

66. Millet, D. B. and Goldstein, A. H.: Evidence of continuing methylchloroform emissions from the United States, Geophys. Res. Lett., 31, L17101, doi:10.1029/2004GL020166, 2004.

67. Mintz, R. and McWhinney, R. D.: Characterization of volatile organic compound emission sources in Fort Saskatchewan, Alberta using principal component analysis, J. Atmos. Chem., 60(1), 83–101, 2008.

68. Montzka, S. A., Trainer, M., Goldan, P. D., Kuster, W. C., and Fehsenfeld, F. C.: Isoprene and its oxidation products, methyl vinyl ketone and methacrolein, in the rural troposphere, J. Geophys. Res., 98(D1), 1101–1111, 1993.

69. Montzka, S. A., Fraser, P. J., Butler, J., Cunnold, D., Daniel, J., Derwent, D., Connell, P., Lal, S., McCulloch, A., Oram, D., Reeves, C., Sanhueza, E., Steele, P., Velders, J. G. M., Weiss, R. F., and Zander, R.: Controlled substances and other source gases, Chapter 1, in Scientific Assessment of Ozone Depletion: 2002, Global Ozone Res. and Monitor. Proj., Vol. No. 47, World Meteorol. Org., Geneva, Switzerland, 2003.

70. Montzka, S. A., Calvert, P., Hall, B. D., Elkins, J. W., Conway, T. J., Tans, P. P., and Sweeney, C.: On the global distribution, seasonality, and budget of atmospheric carbonyl sulfide (COS) and some similarities to CO2, J. Geophys. Res., 112, D09302, doi:10.1029/2006JD007665, 2007.

71. Montzka, S. A., Hall, B. D., and Elkins, J. W.: Accelerated increases observed for hydrochlorofluorocarbons since 2004 in the global atmosphere, Geophys. Res. Lett., 36, L03804, doi:10.1029/2008GL036475, 2009.

72. Niinemets, U¨.: Mild versus severe stress and BVOCs: thresholds, priming and consequences, Trends Plant Sci., 15(3), 145–153, 2010.

73. Nowak, J. B., Davis, D. D., Chen, G., Eisele, F. L., Mauldin III, R. L., Tanner, D. J., Cantrell, C., Kosciuch, E., Bandy, A., Thornton,

D., and Clark, A.: Airborne observations of DMSO, DMS, and OH at marine tropical latitudes, Geophys. Res. Lett., 28(11), 2201–2204, 2001.

74.   O'Doherty, S., Cunnold, D. M., Manning, A., B. R. Miller, B. R., Wang, R. H. J., Krummel, P. B., Fraser, P. J., Simmonds, P. G., McCulloch, A., Weiss, R. F., Salameh, P., Porter, L. W., Prinn, R. G,. Huang, J., Sturrock, G., Ryall, D., Derwent, R. G., and Montzka, S. A.: Rapid growth of hydrofluorocarbon 134a and hydrochlorofluorocarbons 141b, 142b and 22 from Advanced Global Atmospheric Gases Experiment (AGAGE) observations at Cape Grim, Tasmania, and Mace Head, Ireland, J. Geophys. Res., 109, D06310, doi:10.1029/2003JD004277, 2004.

75.   Paatero, P.: Least squares formation of robust non-negative factor analysis, Chemometr. Intell. Lab., 37, 15–35, 1997.

76.   Parrish, D. D., Kuster, W., C., Shao, M., Yokouchi, Y., Kondo, Y., Goldan, P. D., de Gouw, J. A., Koike, M., and Shirai, T.: Comparison of air pollutant emissions among mega-cities, Atmos. Environ., 43(40), 6435–6441, 2009.

77.   Penner, T. J. and Foght, J. M.: Mature fine tailings from oil sands processing harbour diverse methanogenic communities, Can. J. Microbiol., 56(6), 459–470, 2010.

78.   R"ais"anen, T., Ryypp"o, A., and Kellom"aki, S.: Impact of timber felling on the ambient monoterpene concentration of a Scots pine (Pinus sylvestris L.) forest, Atmos. Environ., 42(28), 6759–6766, 2008.

79.   Randerson, J. T., Thompson, M. V., Conway, T. J., Fung, I. Y., and Field, C. B.: The contribution of terrestrial sources and sinks to trends in the seasonal cycle of atmospheric carbon dioxide, Global Biogeochem. Cy., 11, 535–560, 1997.

80.   Randerson, J. T., Field, C. B., Fung, I. Y., and Tans, P. P.: Increases in early season ecosystem uptake explain recent changes in the seasonal cycle of atmospheric at high northern latitudes, Geophys Res. Lett., 26(17), 2765–2768, 1999.

81.   Ras, M. R., Marc´e, R. M., and Borrull, F.: Characterization of ozone precursor volatile organic compounds in urban atmospheres and around the petrochemical industry in the Tarragona region, Sci. Total Environ., 407(14), 4312–4319, 2009.

82. Redeker, K. R., Wang, N. Y., Low, J. C., McMillan, A., Tyler, S. C., and Cicerone, R. J., Emissions of methyl halides and methane from rice paddies, Science, 290, 966–969, doi:10.1126/science.290.5493.966, 2000.

83. Reimann, S., Calanca, P., and Hofer, P.: The anthropogenic contribution to isoprene concentrations in a rural atmosphere, Atmos. Environ., 34(1), 109–115, 2000.

84. Rogers, M. A. and Koons, C. B.: Generation of light hydrocarbons and establishment of normal paraffin preferences in crude oils, Adv. Chem., 103(3), 67–80, 1971.

85. Russo, R. S., Zhou, Y., White, M. L., Mao, H., Talbot, R., and Sive, B. C.: Multi-year (2004–2008) record of nonmethane hydrocarbons and halocarbons in New England: seasonal variations and regional sources, Atmos. Chem. Phys., 10, 4909–4929, doi:10.5194/acp-10-4909-2010, 2010.

86. Ryerson, T. B., Trainer, M., Angevine, W. M., Brock, C. A., Dissly, R. W., Fehsenfeld, F. C., Frost, G. J., Goldan, P. D., Holloway, J. S., Hübler, G., Jakoubek, R. O., Kuster, W. C., Neuman, J. A., Nicks Jr., D. K., Parrish, D. D., Roberts, J. M., Sueper, D. T., Atlas, E. L., Donnelly, S. G., Flocke, F., Fried, A., Potter, W. T., Schauffler, S., Stroud, V., Weinheimer, A. J., Wert, B. P., Wiedinmyer, C., Alvarez, R. J., Banta, R. M., Darby, L. S., and Senff, C. J.: Effect of petrochemical industrial emissions of reactive alkenes and NOx on tropospheric ozone formation in Houston, Texas, J. Geophys. Res., 108(D8), 4249, doi:10.1029/2002JD003070, 2003.

87. Schade, G. W. and Goldstein, A. H.: Increase of monoterpene emissions from a pine plantation as a result of mechanical disturbances, Geophys. Res. Lett., 30(7), 1380, doi:10.1029/2002GL016138, 2003.

88. Sharkey, T. D., Wiberley, A. E., and Donohue, A. R.: Isoprene emission from plants: Why and how, Annal. Botany, 101, 5–18, 2008.

89. Siddique, T., Fedorak, P. M., and Foght, J. M.: Biodegradation of short-chain n-alkanes in oil sands tailings under methanogenic conditions, Environ. Sci. Technol., 40, 5459–5464, 2006.

90. Siddique, T., Fedorak, P. M., MacKinnon, M. D., and Foght, J. M.: Metabolism of BTEX and naphtha compounds to methane in oil sands tailings, Environ. Sci. Technol., 41, 2350–2356, 2007.

91. Siddique, T., Gupta, R., Fedorak, P. M., MacKinnon, M. D., and Foght, J. M.: A first approximation kinetic model to predict methane generation from an oil sands tailings settling basin, Chemosphere, 72, 1573–1580, 2008.

92. Simpson, I. J., Blake, D. R., Rowland, F. S., and Chen, T.-Y.: Implications of the recent fluctuations in the growth rate of tropospheric methane, Geophys. Res. Lett., 29(10), 672–675, 2002.

93. Simpson, I. J., Wang, T., Guo, H., Kwok, Y. H., Flocke, F., Atlas, E., Meinardi, S., Rowland, F. S., and Blake, D. R.: Longterm amospheric measurements of C1–C5 alkyl nitrates in the Pearl River Delta region of southeast China, Atmos. Environ., 40, 1619–1632, 2006.

94. Sive, B. C.: Atmospheric nonmethane hydrocarbons: Analytical methods and estimated hydroxyl radical concentrations, Ph.D Thesis, University of California, Irvine, CA, 1998.

95. Sive, B. C., Varner, R. K., Mao, H., Blake, D. R., Wingenter, O. W., and Talbot, R., A large terrestrial source of methyl iodide, Geophys. Res. Lett., 34, L17808, doi:10.1029/2007GL030528, 2007.

96. Sprengnether, M., Demerjian, K. L., Donahue, N. M., and Anderson, J. G.: Product analysis of the OH oxidation of isoprene and 1,3-butadiene in the presence of NO, J. Geophys. Res., 107(D15), 4268, doi:10.1029/2001JD000716, 2002.

97. Stroud, C., Roberts, J. M., Goldan, P. D., Kuster, W. C., Murphy, P. C., Williams, E. J., Hereid, D., Parrish, D., Sueper, D., Trainer, M., Fehsenfeld, F. C., Apel, E. C., Riemer, D., Wert, B., Henry, B., Fried, A., Martinez-Harder, M., Brune, W. H., Li, G., Xie, H., and Young, V. L.: Isoprene and its oxidation products, methacrolein and methylvinyl ketone, at an urban forested site during the 1999 Southern Oxidants Study, J. Geophys. Res., 106, 8035–8046, 2001.

98. Talbot, R., Mao, H., Scheuer, E., Dibb, J., Avery, M., Browell, E., Sachse, G., Vay, S., Blake, D., Huey, G., and Fuelberg, H.: Factors influencing the large-scale distribution of $Hg_$ in the Mexico City area and over the North Pacific, Atmos. Chem. Phys., 8, 2103–2114, doi:10.5194/acp-8-2103-2008, 2008.

99.   Thurston, G. D. and Spengler, J. D.: A quantitative assessment of source contributions to inhalable particulate matter pollution in metropolitan Boston, Atmos. Environ., 19(1), 9–25, 1985.

100. Timoney, K. P. and Lee, P.: Does the Alberta tar sands industry pollute? The scientific evidence, Open Conservation Biol. J., 3, 65–81, 2009.

101. Vay, S. A., Anderson, B. E., Conway, T. J., Sachse, G. W., Collins, J. E., Blake, D. R., andWestberg, D. J.: Airborne observations of the tropospheric CO_2 distribution and its controlling factors over the South Pacific Basin, J. Geophys. Res., 104(D5), 5663–5676, 1999.

102. Vay, S. A., Woo, J. H., Anderson, B. E., Thornhill, K. L., Blake, D. R., Westberg, D. J., Kiley, C. M., Avery, M. A., Sachse, G. W., Streets, D. G., Tsutsumi, Y., and Nolf, S. R.: The influence of regional-scale anthropogenic emissions on CO_2 distributions over the western North Pacific, J. Geophys. Res., 108(D20), 8801, doi:10.1029/2002JD003094, 2003.

103. Warneke, C., McKeen, S. A., de Gouw, J. A., Goldan, P. D, Kuster, W. C., Holloway, J. S., Williams, E. J., Lerner, B. M., Parrish, D. D., Trainer, M., Fehsenfeld, F. C., Kato, S., Atlas, E. L., Baker, A., and Blake, D. R.: Determination of urban volatile organic compound emission ratios and comparison with an emissions database, J. Geophys. Res., 112, D10S47, doi:10.1029/2006JD007930, 2007.

104. Warneke, C., de Gouw, J. A., Del Negro, L., Brioude, J., McKeen, S., Stark, H., Kuster, W. C., Goldan, P. D., Trainer, M., Fehsenfeld, F. C., Wiedinmyer, C., Guenther, A. B., Hansel, A., Wisthaler, A., Atlas, E., Holloway, J. S., Ryerson, T. B., Peischl, J., Huey, L. G., and Case Hanks, A. T.: Biogenic emission measurement and inventories determination of biogenic emissions in the eastern United States and Texas and comparison with biogenic emission inventories, J. Geophys. Res., 115, D00F18, doi:10.1029/2009JD012445, 2010.

105. Watson, J. G., Chow, J. C., and Fujita, E. M.: Review of volatile organic compound source apportionment by chemical mass balance, Atmos. Environ., 35(9), 1567–1584, 2001.

106. Watts. S. F.: The mass budgets of carbonyl sulfide, dimethyl sulfide, carbon disulfide and hydrogen sulfide, Atmos. Environ., 34, 761– 779, 2000.

107. Weinheimer, A. J., Walega, J. G., Ridley, B. A., Gary, B. L., Blake, D. R., Blake, N. J., Rowland, F. S., Sachse, G. W., Anderson, B. E., and Collins, J. E.: Meridional distributions of NOx, NOy, and other species in the lower stratosphere and upper troposphere during AASE II, Geophys. Res. Lett., 21, 2583–2586, 1994.

108. White, M. L., Russo, R. S., Zhou, Y., Mao, H., Varner, R. K., Ambrose, J., Veres, P., Wingenter, O. W., Haase, K., Stutz, J., Talbot, R., and Sive, B. C.: Volatile organic compounds in northern New England marine and continental environments during the ICARTT 2004 campaign, J. Geophys. Res., 113, D08S90, doi:10.1029/2007jd009161, 2008.

109. White, M. L., Russo, R. S., Zhou, Y., Ambrose, J. L., Haase, K., Frinak, E. K., Varner, R. K., Wingenter, O. W., Mao, H., Talbot, R., and Sive, B. C.: Are biogenic emissions a significant source of summertime atmospheric toluene in the rural Northeastern United States?, Atmos. Chem. Phys., 9, 81–92, doi:10.5194/acp-9-81-2009, 2009.

110. White, M. L., Zhou, Y., Russo, R. S., Mao, H., Talbot, R., Varner, R. K., and Sive, B. C.: Carbonyl sulfide exchange in a temperate loblolly pine forest grown under ambient and elevated CO_2, Atmos. Chem. Phys., 10, 547–561, doi:10.5194/acp-10-547-2010, 2010

111. Xiao, Y., Logan, J. A., Jacob, D. J., Hudman, R. C., Yantosca, R., and Blake, D. R.: Global budget of ethane and regional constraints on U.S. sources, J. Geophys. Res., 113, D21306, doi:10.1029/2007JD009415, 2008.

112. Yokouchi, Y., Osada, K., Wada, M., Hasebe, F., Agama, M., Murakami, R., Mukai, H., Nojiri, Y., Inuzuka, Y., Toom-Sauntry, D., and Fraser, P.: Global distribution and seasonal concentration change of methyl iodide in the atmosphere, J. Geophys. Res., 113, D18311, doi:10.1029/2008JD009861, 2008.

113. Zhang, Y. L., Guo, H., Wang, X. M., Simpson, I. J., Barletta, B., Blake, D. R., Meinardi, S., Rowland, F. S., Cheng, H. R., Saunders, S. M., and Lam, S. H. M.: Emission patterns and spatiotemporal variations of halocarbons in the Pearl River Delta

region, southern China, J. Geophys. Res., 115, D15309, doi: 10.1029/2009JD013726, 2010.

Citations

CHAPTER 1

J Meldrum, S Nettles-Anderson, G Heath, and J Macknick, Life Cycle Water Use for Electricity Generation: A Review and Harmonization of Literature Estimates, doi:10.1088/1748-9326/8/1/015031.

CHAPTER 2

Marco Schubert, Bill S. Hansson, and Silke Sachse, The Banana Code—Natural Blend Processing in the Olfactory Circuitry of Drosophila Melanogaster, doi: 10.3389/fphys.2014.00059.

CHAPTER 3

Sabtanti Harimurti, Anisa Ur Rahmah, Abdul A. Omar and Thanapalan Murugesan, 2014. Effect of Bicarbonate on the Mineralization of Methyldiethanolamine by using UV/H2O2. Journal of Applied Sciences, 14: 1147-1153.

CHAPTER 4

J. R. Roscioli, T. I. Yacovitch, C. Floerchinger, A. L. Mitchell, D. S. Tkacik, R. Subramanian, D. M. Martinez, T. L. Vaughn, L. Williams5, D. Zimmerle, A. L. Robinson, S. C. Herndon, and A. J. Marchese, Measurements of Methane Emissions from Natural Gas Gathering Facilities and Processing Plants: Measurement Methods, doi:10.5194/amtd-7-12357-2014.

CHAPTER 5

A. Nasir, P. Pilidis, S. Ogaji, H. T. Abdulkarim, and A. El-Suleiman, Electric Motor Drive for Natural Gas Compression in Pipeline: Techno-economic Analysis, ISSN: 2319-8753.

CHAPTER 6

Aviva Litovitz, Aimee Curtright, Shmuel Abramzon, Nicholas Burger, and Constantine Samaras, Estimation of Regional Air-Quality Damages from Marcellus Shale Natural Gas Extraction in Pennsylvania, doi:10.1088/1748-9326/8/1/014017.

CHAPTER 7

Alîne Doherty, Eilín Walsh, Kevin P. McDonnell, the Direct Use of Post-Processing Wood Dust in Gas Turbines, doi.org/10.4236/jsbs.2012.23009

CHAPTER 8

I. J. Simpson, N. J. Blake, B. Barletta, G. S. Diskin, H. E. Fuelberg, K. Gorham, L. G. Huey, S. Meinardi, F. S. Rowland, S. A. Vay, A. J. Weinheimer, M. Yang, and D. R. Blake, Characterization of Trace Gases Measured Over Alberta Oil Sands Mining Operations: 76 Speciated C2–C10 Volatile Organic Compounds (VOCs), CO2, CH4, CO, NO, NO2, Noy, O3 and SO2, doi:10.5194/acp-10-11931-2010.

Index